Next

Transmaterial Next

> A Catalog of Materials
> That Redefine Our Future

EDITED BY
Blaine Brownell

Princeton
Architectural
Press

New York

Contents

1 Concrete

2 Mineral

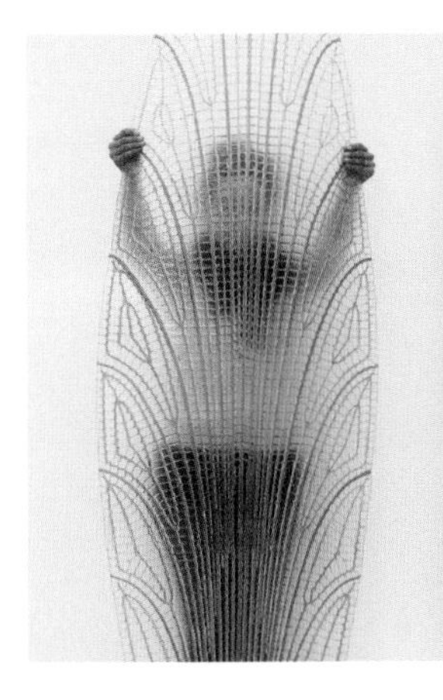

3 Metal

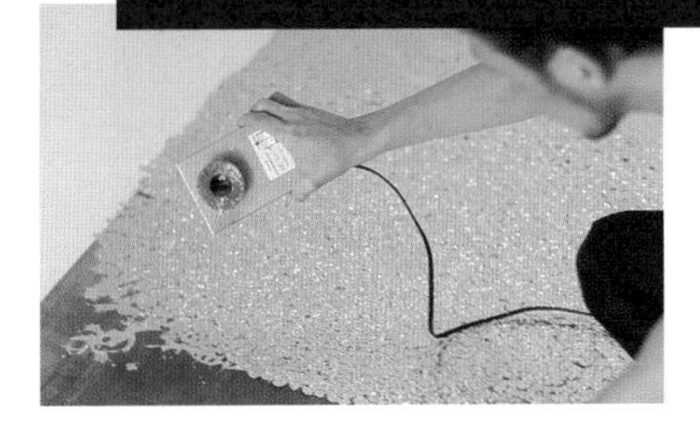

4 Wood and Biomaterials

5 Plastic and Rubber

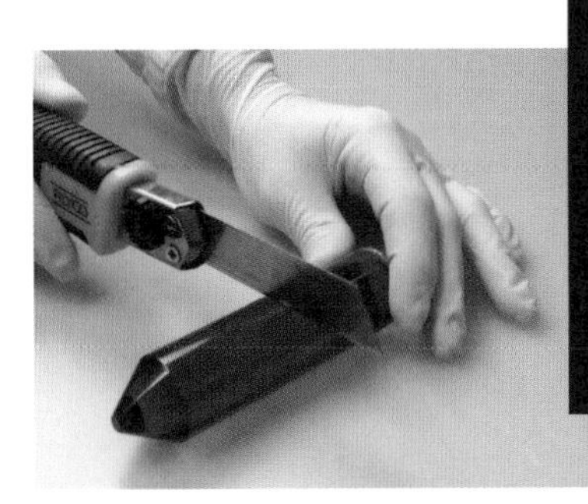

6 Glass

7 Paint and Coatings

8 Fabric

9 Light

10 Digital

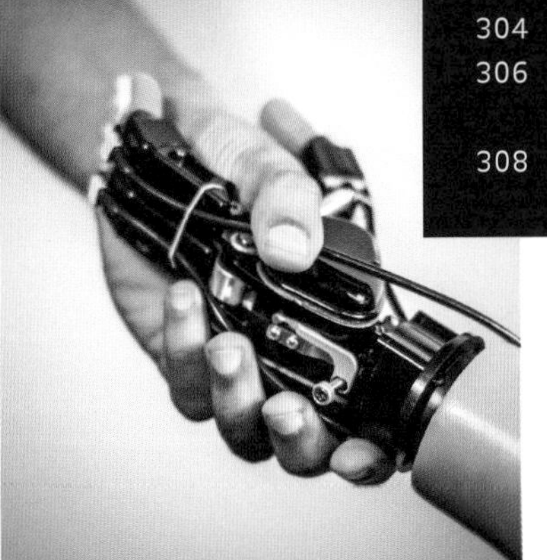

INDEXES

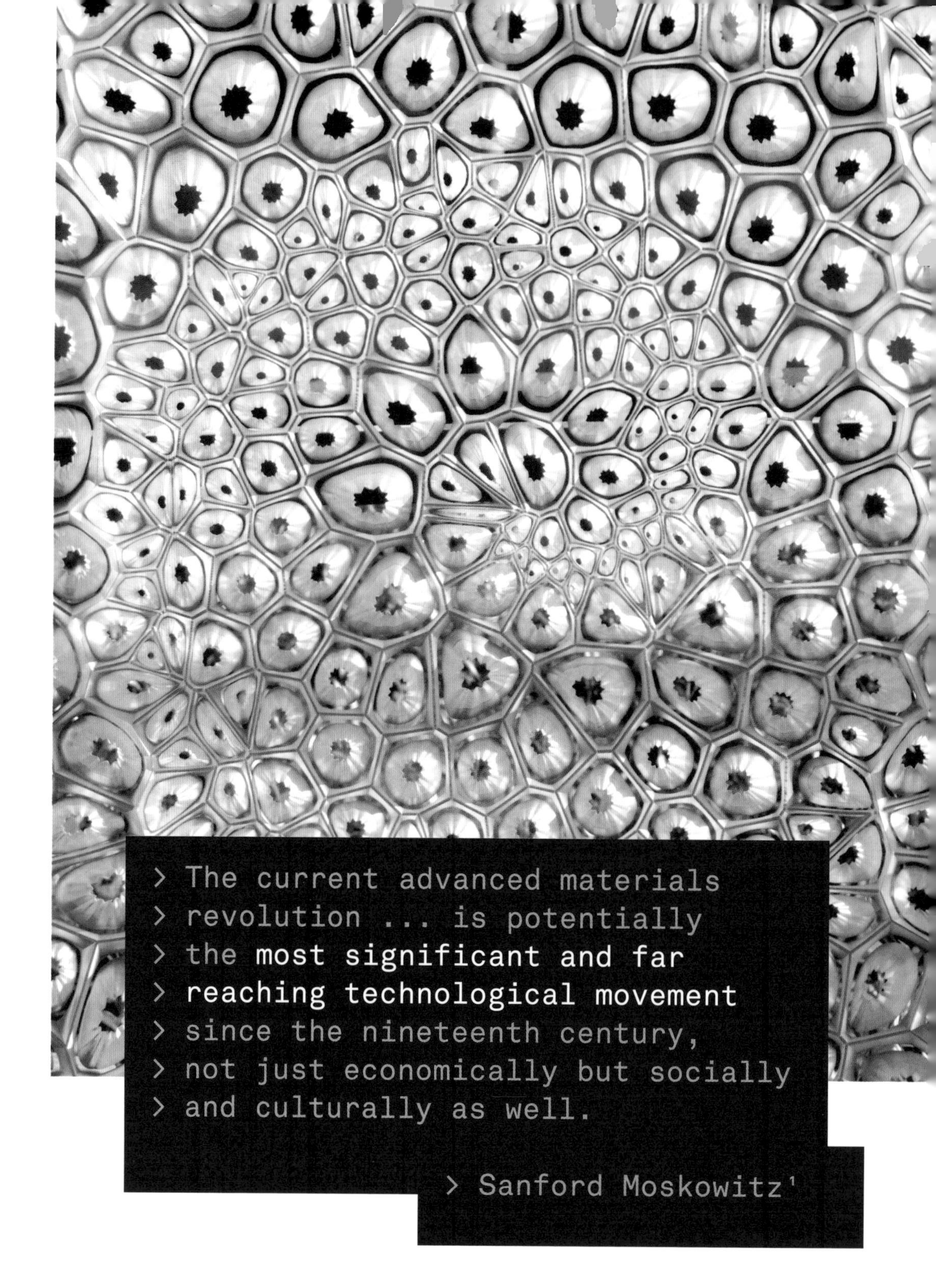

> The current advanced materials
> revolution ... is potentially
> the **most significant and far**
> **reaching technological movement**
> since the nineteenth century,
> not just economically but socially
> and culturally as well.

> Sanford Moskowitz[1]

> Introduction

Toward the end of the twentieth century, technology expert Tom Forester predicted that three megatechnologies would come to dominate global industrial activity: biotechnology, information technology, and new materials.[2] Today an international material revolution is underway, motivated by heightened material research investigations, in addition to an intense sociocultural focus on the expressive possibilities of material applications.

Given the fundamental effect that material choices have on the flows of resources, emissions, and waste, designers and manufacturers have realized how crucial it is to increase their material expertise in order to make more informed decisions about the future built environment. New material technologies now exert an influence on all industrial sectors, shaping building codes, construction standards, trade organization recommended practices, and environmental rating systems. The extent and pace of change in material-based industries are unprecedented, and this change will bring new opportunities and challenges.

Despite the explosion of new material offerings and the significance of material innovation in the design and construction fields, there is a lack of knowledge about the comprehensive scope of this trend and its future implications. Various organizations track material trends according to areas of specialization. For example, the U.S. Geological Survey monitors raw materials and ores for industry executives and market analysis. Sustainability-focused organizations, such as BuildingGreen, provide updates about environmentally responsible material choices to architects and builders. Meanwhile, online research portals, like MaterialsViews, share advances in material experiments with scientific audiences.[3] As a result, broad material trajectories with connections to many fields are poorly understood.

There is also a lack of understanding of the opportunities provided by new materials and the potential for new material applications to transform architecture and design practice. From the early writings of Pliny and Vitruvius to the manuals published by contemporary product associations, the emphasis in materials literature has been on existing demonstrated practices—despite the fact that the rote following of conventional methods precludes the development of new ideas in the designed environment.

This book takes a cross-sectional look across a variety of industries in an effort to identify the most important material opportunities for architecture and design—as well as their most significant implications for the future of society. Resource trends, environmental goals, scientific achievement, and design applicability are all relevant to questions about tomorrow's physical environment, and this book focuses on next-generation materials because of the huge changes they are anticipated to bring.

DRIVERS

Material innovation is fueled by many factors. For simplicity's sake, we could try to distinguish these in terms of internal and external drivers. Internal drivers pertain to the motivations and creative insights of a particular research team or manufacturer. External drivers concern outside forces, such as industry goals or environmental objectives. In reality, these drivers are often difficult to distinguish, as individuals cannot be separated from outside influences. In fact, this mutual reinforcement of internal and external forces is intensifying, based on the increased efficacy of the internet and global telecommunications. For example, a wood scientist's inspiration to create a successful ➔ **wood foam insulation** may come from adverse environmental reports about the standard petroleum-based product, coupled with news about advances in previously manufactured foamed materials, such as metals and ceramics (see "Wood Foam," page 126). In this way, new materials can develop out of divergent influences, such as market concerns and technology transfer methods.

According to economist Sanford Moskowitz, there have been three distinct periods of new-materials development since the end of the nineteenth century.[4] The first occurred between 1880 and 1930, and focused on the mass production of steel and other metals, in addition to the development of coal-tar products. The second transpired between 1930 and the 1970s, with the shift from coal to petroleum and the surge of oil-derived synthetic materials. The third period, which emerged in the 1980s and is now exerting a broad impact, saw the introduction of advanced materials made from both hydrocarbon- and carbohydrate-based resources. As defined by the Versailles Project on Advanced Materials and Standards (VAMAS), which was initiated within the framework of the 1982 Economic Summit of the Heads of State of the G7 countries, advanced materials invited a separate classification based on their "special properties" and "novel or advanced processing procedures."[5]

In this book, I have interpreted VAMAS criteria liberally to provide a broader spectrum of the scientific and artistic innovation currently underway. We can

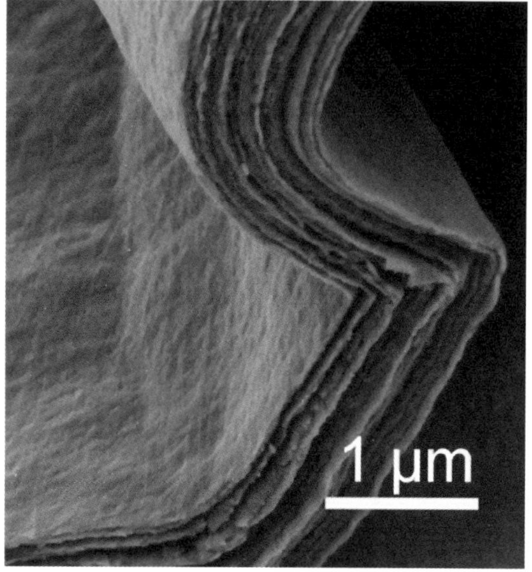

evaluate the correlative external and internal drivers influencing material innovation according to four categories: science, technology, culture, and the environment.

SCIENCE

As much as the laboratory is a place of quiet, methodical concentration, it is also a war room. Researchers are under intense pressure to generate new knowledge and to publish it before their competition does. Scientific institutions also compete fiercely for superior rankings, more grant funding, and better faculty and students. All of this drive has led to what some scientists call the "golden age" for materials.[6] The Latin motto *citius, altius, fortius* (faster, higher, stronger) embodies the competitive edge conferred by new laboratory-generated supermaterials, such as Aerographene or metallic microlattice, which are two of the lightest substances ever made, or nanocellulose composites, wood fibers processed to exhibit the strength of Kevlar. Nanotechnology and unexpected material mash-ups have delivered mutant materials, such as ↙ **ceramic paper**, spinel-based transparent armor windows, and bendable concrete—substances with properties previously found in two or more different materials. Multiple functions are also increasingly combined in material technologies with augmented capacities, such as DysCrete, a power-harvesting concrete; or a switchable OLED (Organic Light Emitting Diode) panel that can act as a transparent window or an illuminated plane. As BASF designer Alex Horisberger claims, "Tomorrow's materials need to perform on multiple levels."[7]

TECHNOLOGY

Technology migrates scientific competition to the marketplace. The arduous process of developing ideas into commercial products is fraught with complex challenges, and how effectively manufacturers can deliver successful ideas to consumers can be a matter of economic survival. Over time, technology and economic growth have become more tightly interconnected. According to Moskowitz, "Today, a new generation of materials plays a far greater role in determining industrial competitiveness than their counterparts did in the past."[8] This competitiveness results in a strong push toward automation in various forms. Robotics technologies continue to experience strong growth, and they will increasingly be used to construct buildings, in addition to smaller products and vehicles. Examples include robots that construct bridges out of steel, flying robots that build towers of lightweight bricks, and ↓ **Facadeprinter robots that paint building envelopes**. 3D printing is similarly expanding in scope and scale, adding capabilities to print materials such as transparent glass and to print earth at the size of a building, respectively. Self-assembly represents another frontier for automation, with time-based 4D printing structures and modular robots that build their own stacking frameworks.

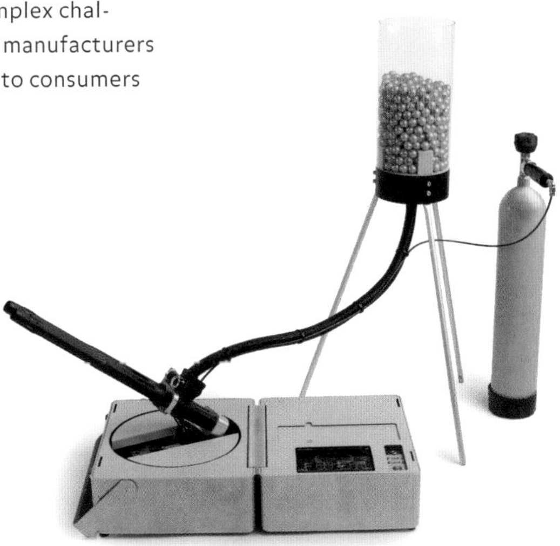

CULTURE

In 2012 3D Robotics CEO Chris Anderson claimed that "The past ten years have been about discovering new ways to create, invent, and work together on the Web. The next ten years will be about applying those lessons to the real world." [9] The most significant aspect of the maker movement is not what is being made, but who is making it. The proliferation of inexpensive CAD software and digital fabrication tools has enabled a large population to design and manufacture its own products and assemblies. This distributed creative platform has inspired consumer audiences who are weary of mass-produced goods and who wish to support fresh creations made by independent artists. According to branding consultancy thefuturelaboratory.com editor Jonathan Openshaw, large manufacturers are beginning to connect with cottage industries to gain competitive advantage, thus indicating that "materials innovations previously dismissed as art-school projects could be big business." [10] This reconnection between craft and craftsperson has also entered the realm of materials themselves, in the form of DIY materials like vessels made from flour, agricultural waste, and limestone; or ↗ **a process that transforms salvaged rubble into new masonry units**.[11] Even for nonmakers, the DIY crusade promotes a deeper connection between materials and their applications, and invites designs that encourage tactility and interactivity.

ENVIRONMENT

Manufacturers and consumers are increasingly privileging environmental considerations in their material choices. According to Openshaw, "Sustainability is being taken to the heart of industry—not as a publicity stunt but as a solid business strategy." [12] Environmental priorities have inspired a variety of innovations, each targeted at a particular opportunity or concern. For example, renewable energy harvesting is becoming more deeply integrated into the built environment, with technologies like transparent solar windows and self-illuminating, electric

car–charging Solar Roadways. Embodied energy reduction in manufacturing has motivated the development of products like portland cement–free concrete block or flash-formed iron, a gas-reduction process that bypasses the traditional coke oven. Resource concerns have inspired the recapture of waste materials, such as disposable plastic bags or marine plastic, for making new products. Furthermore, bioreceptive products like Biological Concrete and 12 Blocks encourage wildlife habitat and plant growth on buildings.

Sustainability is closely aligned with the concepts of resilience and regenerative design, which often find inspiration in biological and ecological models. For example, self-diagnosing and self-repairing materials employ smart properties to enhance longevity—such as a smart bandage that warns of infection or self-healing polymer that can re-form into a single object after being completely divided in two. Researchers are creating adaptive, responsive technologies that use simple material strategies to substitute for more complex mechanical systems, such as a wood window that opens and closes automatically based on relative humidity levels, or a fiber-reinforced composite lamella that exhibits dynamic properties like nonautonomous mechanics in particular plants. As design innovator Eben Bayer has said, "Biology represents the world's greatest technology."[13] Inspiration from nature has also motivated the quickly expanding fields of biodesign and bioengineering, in which researchers partner with natural organisms to create

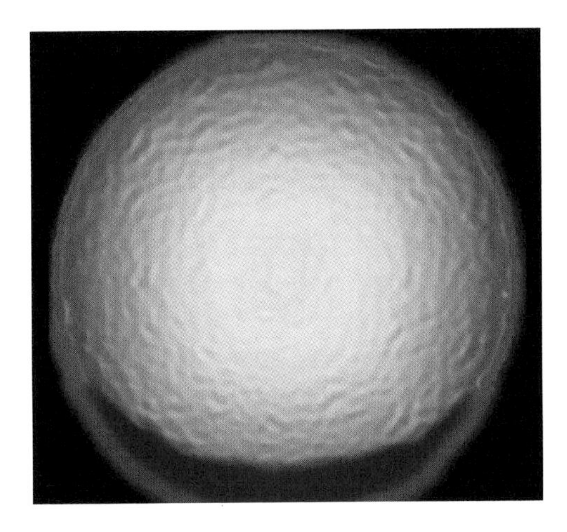

new products. Examples include bricks grown like coral using bacteria, bioengineered spider's thread for making high-performance textiles, and ↑ **a bioluminescent microbe-powered lightbulb**. The more intimate these cooperative partnerships with nature become, the more human technology will resemble biology itself. As designers Koert van Mensvoort and Hendrik-Jan Grievink declare in a clever revision to Arthur C. Clarke's familiar utterance, "Any sufficiently advanced technology is indistinguishable from nature."[14]

THE SCIENCE OF NEXT

Accelerating change and increased volatility in financial markets, weather systems, and other large forces have encouraged the development of the predictive sciences. In 2008 the U.S. National Nuclear Security Administration launched a multiyear predictive science program in collaboration with several universities to apply "verified and validated computational simulations to predict properties of complex systems," including biological systems, climate modeling, and efficient manufacturing.[15] From a resource perspective, complex computational analysis is frequently employed by a variety of industries to analyze known material reserves and make predictions about the future resource base. With increased access to material supply chain data, it is possible to anticipate future quandaries, such as resource criticality or resource exhaustion. For example, geological engineers and energy analysts continually monitor critical indicators for mineral resources, such as the rate of discovery, rate of production, market price, and ore quality.

When it comes to material innovation, however, predictive science gets fuzzy. In *What Technology Wants*, author Kevin Kelly presents an evolutionary model of technology as an *exotropic* system, or an ordered force with sustained flow, that exhibits long-term trajectories but is unpredictable in the short term: "Technology's imperative is not a tyrant ordering our lives in lock-step. Its inevitabilities are not scheduled prophecies. They are more like water behind a wall, an incredibly strong urge pent up and waiting to be released."[16] Within this long-term arc, innovation is even harder to predict, in part because we still struggle to define it. "We have no shortage of theories to instruct us how to make our organizations more creative, or explain why tropical rain forests engineer so much molecular diversity," author Steven Johnson explains. "What we lack is a unified theory that describes the common attributes shared by all those innovation systems." [17]

Economist Clayton M. Christensen tracks innovation in terms of *disruptive technology*, whereby a sufficiently novel and advantageous technology unexpectedly displaces established models in the marketplace. Examples include the telephone, which displaced the telegraph; or the personal computer, which displaced the typewriter. In *Seeing What's Next*, Christensen and co-authors Scott D. Anthony and Erik A. Roth apply a customer-centered model for presaging future disruptive technologies, arguing that "predicting whether disruptive innovations are taking root, and predicting how they will affect the mainstream of a market in the future, requires watching the low end, new markets, and new contexts." [18] Critics argue that disruptive innovation can equally originate at the top end of the market, as seen in products made by Apple, and that Christensen's definition is too narrowly focused.[19] For our purposes in anticipating potential changes in the physical environment, we can apply the notion of disruption in general terms. For example, one could argue that LED technology, which has rapidly overtaken incandescent and fluorescent lighting, is a disruptive innovation. Other technologies, such as electrochromic windows, are relatively recent entrants to the market and would therefore not earn a disruptive label for Christensen; however, industry analysts now generally use this lens to make future predictions. For example, a recent McKinsey "disruptive technology" report on compressorless air-conditioning and electrochromic windows states, "Today, these technologies are expensive, but by 2020, they could begin to cost only about half as much to install as current state-of-the-art cooling and window technologies." [20] Such predictions are based on technological capacities—which include future improvements anticipated by researchers and manufacturers—as well as their viability in the marketplace.

ABOUT THE BOOK

Transmaterial Next presents, by definition, such a future-oriented focus. The book consists of more than a hundred materials that have significant potential to transform future products, buildings, and cities. In some cases this influence is specific to a particular technology, such as solar-harvesting windows. However, in many instances, the material entries represent broader trends that are significant. For example, the ➚ **Artichair, which is made of artichoke thistle fiber–reinforced polymer**, does not presage a commercial run on artichoke thistles—but rather a consequential trend toward incorporating more alternative biomaterials in furniture design and manufacturing (see page 106). In any case, the anticipated influence is identified under "Future Impact" for each material entry.

As with previous volumes of the Transmaterial series, I have taken an intentionally broad view of what constitutes "material," and I include a wide variety of material-based products, assemblies, processes, and applications. I even purposefully push the limits of the definition by including chapters on light and digital technologies, since they require materials to function although they are commonly understood as having immaterial properties. The rationale is that an overly rigid focus on materials is better suited to an engineering textbook, and that it would prevent the inclusion of significant creative applications that represent unexpected material combinations and methodologies for design. In this sense,

the book is not just about materials; it is about ideas. It is for this reason that I have chosen to focus on the next generation of materials, which fires the imagination and invites rumination about the imminent changes in design and architecture.

Another reason for focusing on what's next is that the coming wave of material technologies promises unparalleled change in terms of quantity and quality. Moskowitz states that "on a very fundamental level, the sheer number and variety of new materials currently in play or on the horizon is unprecedented, even compared to the technically active post World War II decades" and that "over time, we note an increasing flexibility and range of innovation." [21] Furthermore, the current third wave of material development offers something entirely new based on flexible inorganic-organic hybridization, "thus adding a third dimension to materials innovation not possible in earlier periods." [22]

Because a book has a longer anticipated shelf life than a news article, two timescale considerations of "next" operate here. The first timescale literally addresses material technologies that are predicted to exert important influences by 2020. By contrast, the second timescale has a longer-term outlook. Discussions of bioengineered materials, the carbohydrate economy, or enchanted systems of objects, for example, address changes that have far-flung temporal implications. These will remain important forces for decades, if not centuries, to come. Regardless of timescale, all "next" materials meet three primary criteria: they exhibit transformative potential over conventional versions, they physically exist in some form (i.e., are not just renderings), and they are on a path toward commercialization (or have just recently been commercialized). It is important that all of the examples have some physical manifestation (i.e., not just renderings), although they can be not yet widely available. Development is represented by a commercial readiness scale, with a range between one and five (one represents an early prototype, while five indicates full-scale mass production for the public market). Some authors do not intend to commercialize their works; in these cases, a rating is established based on a commensurate level of development.

Like the other *Transmaterial* volumes, the book addresses a wide range of creative production, from the arts to the sciences and from advanced laboratories to amateur workshops. It is organized according to familiar material categories, such as concrete, metal, and glass. These categories are intentionally general in nature, based on how designers and specifiers think about materials in simple terms. The motivation here is design application rather than scientific classification; thus, I have purposefully kept concrete, glass, and mineral as separate chapters, despite their shared material traits. I have also avoided a separate category for composites, since so many new materials fit this definition, thereby reducing this classification's distinctiveness. The categories follow an order that is loosely based on the MasterFormat system developed by the Construction Specifications Institute, and the materials within each chapter are listed alphabetically.[23] Each chapter opens with a broad overview of the transformations underway in each category, in addition to a review of opportunities and challenges.

TRENDS

As established in the first *Transmaterial* volume, several broad categories serve to elucidate significant material transformations. These classifications highlight important themes shared by dissimilar products and make unexpected connections. For example, an aluminum floor system and polypropylene chair are made of different substances, but they could be similarly notable in their use of recycled materials. The seven broad categories I have used are as follows:

ULTRAPERFORMING

Throughout history, material innovation has been defined by the persistent testing of limits. A lexical shortening of *ultra-high-performing*, *ultraperforming* describes materials that are stronger, lighter, more durable, and more flexible than their conventional counterparts. These materials are significant because they shatter known boundaries and necessitate new thinking about the shaping of our physical environment.

The ongoing pursuit of thinner, more porous, and less opaque products indicates a notable movement toward greater exposure and ephemerality. It is no surprise that ultraperforming materials are often expensive and difficult to obtain, although many of these products eventually reach a broad market.

MULTIDIMENSIONAL

Materials are physically defined by three dimensions, but many products have long been conceived as flat surfaces. One trend exploits the *z* axis in the manufacture of a wide variety of materials for various uses, ranging from fabrics to wall and ceiling treatments. Greater depth allows thin materials to become more structurally stable, and materials with enhanced texture and richness are often more visually impressive. Augmented dimensionality will likely continue to be a growing movement, especially considering the technological trends toward miniaturization, systems integration, and prefabrication.

REPURPOSED

Repurposed materials may be defined as surrogates, or materials that are used in the place of materials conventionally used in an application. Repurposed materials provide several benefits, such as replacing precious raw materials with less endangered, more plentiful ones; diverting products from the waste stream; implementing less toxic manufacturing processes; and defying convention. A subset of this group comprises objects considered repurposed in terms of their functionality, such as tables that become light sources and art that becomes furniture.

As a trend, repurposing underscores the desire for adaptability and an increasing awareness of our limited resources. While the performance of repurposed materials is not always identical to that of the products they replace, sometimes new and unexpected benefits arise from their use.

RECOMBINANT

Recombinant materials consist of two or more different materials that act in harmony to create a product whose performance is greater than the sum of its parts. Such hybrids are created when inexpensive or recyclable products are used as filler, when a combination allows for the achievement of multiple functions, when a precious resource may be emulated by combining less-precious materials, or when different materials act in symbiosis to exhibit high-performance characteristics.

Recombinant materials have long proved their performance in the construction industry. Reinforced concrete, which benefits from the compressive strength and fireproof qualities of concrete and the tensile strength of steel, is a classic recombination. These materials often consist of downcycled components, which may be difficult if not impossible to reextract, and the success of recombinant materials is based on their reliable integration, which is not always predictable. However, the continued value exhibited by many such hybrids is reason for recombinant materials' growing popularity.

INTELLIGENT

Intelligent is a catchall term for materials that are designed to improve their environment and that often take inspiration from biological systems. They can act actively or passively and can be high- or low-tech in character. Many materials in this category indicate a focus on the manipulation of the microscopic scale.

Intelligence is not used here to describe products that have sentient capabilities but rather products that are inherently smart by design. The varied list of benefits provided by these materials includes pollution reduction, water purification, solar-radiation control, natural ventilation, and power generation. An intelligent product may simply be a flexible or modular system that adds value throughout its life cycle.

TRANSFORMATIONAL

Transformational materials undergo a physical metamorphosis based on environmental stimuli. This change may occur automatically based on the inherent properties of the material, or it may be user driven.

Like intelligent materials, transformational materials provide a variety of benefits, including waste reduction, enhanced ergonomics, solar control, and illumination, as well as unique phenomenological effects. Transformational products offer multiple functions rather than a single use, provide benefits that few might have imagined, and help us view the world differently.

INTERFACIAL

The interface has been a popular design focus since the birth of the digital age. Interfacial materials, products, and systems navigate between the physical and virtual realms. As we spend greater amounts of time interacting with computer-based tools and environments, the bridges that facilitate the interaction between the two worlds are subject to further scrutiny.

So-called interfacial products may be virtual instruments that control material manufacture or physical manifestations of digital fabrications. These tools provide novel capabilities such as enhanced technology-infused work environments, rapid prototyping of complex shapes, integration of digital imagery within physical objects, and making the invisible visible.

Interfacial materials employ the latest computing and communications technologies and suggest future trajectories for society. Interfacial materials are not infallible, but they expand our capabilities into uncharted territory.

LOOKING AHEAD

This book is dedicated to the thoughtful architects, designers, engineers, and other creative professionals who are passionate about materials and the new opportunities they create within our physical environment—as well as design

NAME
The trademarked or colloquial name of the particular entry being featured

DESCRIPTION
A brief, generic explanation of each entry

SUMMARY
A basic text description of each entry

ADDITIONAL DATA
The following information is also used to describe product entries: contents, applications, types or sizes, environmental benefits, industry tests or examinations, limitations, and manufacturer contact information.

FUTURE IMPACT (NEW)
Description of the transformative influence anticipated for each entry

COMMERCIAL READINESS (NEW)
A five-point rating system that quantifies the level of product development:
1 = early prototype, 5 = fully commercialized

PRODUCT

CEMENT-FREE CONCRETE MASONRY UNIT
Watershed Block

Watershed Block is an ecologically responsive and attractive replacement for the traditional concrete block (CMU), one of the most commonly used building materials. Watershed Materials, a technology start-up in Napa, California, supported by the National Science Foundation, began development of Watershed Block in 2011. The company reduces the cement in structural masonry with an ultra-high-compaction manufacturing process that mimics the way stone is formed in nature, along with novel designs that take advantage of the aluminosilicates found in aggregates around the world.

Watershed Materials initially reduced the cement by 50 percent, but further materials and manufacturing refinements enabled the complete elimination of cement. The manufacturer accomplishes the removal of cement with several mix designs that explore the geopolymerization of natural aluminosilicates and the reuse of industrial waste products to form unique binding combinations.

CONTENTS
85–95 percent locally sourced recycled aggregates; 5–15 percent cement-free binders, including geopolymers and lime / slag / aluminosilicate blend, or 5–10 percent ordinary cement

APPLICATIONS
Weight-bearing structural wall systems (same as traditional concrete block)

TYPES / SIZES
Same as traditional concrete block (CMU):
4 × 8 × 16" (10 × 20 × 41 cm)
6 × 8 × 16" (15 × 20 × 41 cm)
8 × 8 × 16" (20 × 20 × 41 cm)

ENVIRONMENTAL
Cement reductions between 50 percent and 100 percent compared with traditional CMU, reductions in material transportation and water consumption

TESTS / EXAMINATIONS
ASTM C90

LIMITATIONS
Not suitable for areas with high freeze–thaw cycles

FUTURE IMPACT
Reduction or elimination of energy-intensive cement from the ubiquitous concrete block

COMMERCIAL READINESS
● ● ● ● ●

CONTACT
Watershed Materials
11 Basalt Road, Napa, CA 94558
707-224-2532
www.watershedmaterials.com
info@watershedmaterials.com

REPURPOSED

46 Concrete

042223-003 47

CLASS
Defines each entry as one of the following:
- Material
- Product
- Process

CATEGORY
Refers to the basic materiality of the product:
- Concrete
- Mineral
- Metal
- Wood and Biomaterials
- Plastic and Rubber
- Glass
- Paint and Coatings
- Fabric
- Light
- Digital

ID
This nine-digit identification number is unique to each entry.

TREND
This field assigns one of the seven trends mentioned in the introduction to each entry:
- Ultraperforming
- Multidimensional
- Repurposed
- Recombinant
- Intelligent
- Transformational
- Interfacial

students and material enthusiasts who would like to learn more about new material technologies and opportunities for innovation. The book invites consideration of next-generation materials not only from a mechanical point of view, but also from a broader theoretical perspective. In the creativity-focused blog brainpickings.org, author Maria Popova states, "In an age obsessed with practicality, productivity, and efficiency, I frequently worry that we are leaving little room for abstract knowledge and for the kind of curiosity that invites just enough serendipity to allow for the discovery of ideas we didn't know we were interested in until we are, ideas that we may later transform into new combinations with applications both practical and metaphysical."[24] At a time when society is increasingly turning to expedient, quantitative, return-on-investment-focused solutions to contemporary problems, the thoughtful consideration of the ways in which we evaluate, design, construct, and transform the physical environment is more important than ever. The change is too swift, and the consequences too great, to merely maintain status quo practices without deeper consideration of their broader implications and possibilities. It is my hope that this book will enable you to engage these implications and opportunities in a new way, and that it will inspire your own creative endeavors.

For more information, I invite you to peruse the website transmaterial.net, where I will continue to share materials and applications that exhibit the potential to transform the future physical environment.

---

1 Sanford L. Moskowitz, *The Advanced Materials Revolution: Technology and Economic Growth in the Age of Globalization* (New York: Wiley, 2009), 3.
2 Tom Forester, ed., *The Materials Revolution: Superconductors, New Materials, and the Japanese Challenge* (Cambridge, MA: MIT Press, 1988).
3 See accessed January 24, 2016, http://www.usgs.gov, http://www2.buildinggreen.com, and http://www.materialsviews.com.
4 Moskowitz, *Advanced Materials Revolution*, 11.
5 James G. Early and Harry L. Rook, "Versailles Project on Advanced Materials and Standards (VAMAS)," *Advanced Materials* 8, no. 1 (1996): 9–10.
6 "Material Difference," *Economist Technology Quarterly* (December 5, 2015): 3.
7 Alex Horisberger, "Make Matter Match," *Frame* 107 (November 2, 2015): 170.
8 Moskowitz, *Advanced Materials Revolution*, 18.
9 Chris Anderson, *Makers: The New Industrial Revolution* (New York: Crown Business, 2012), 17.
10 Jonathan Openshaw, "Touchy Subject," *Frame* 107 (November 2, 2015): 160.
11 Valentina Rognoli, Massimo Bianchini, Stefano Maffei, and Elvin Karana, "DIY Materials," *Materials and Design* 86 (2015): 693–95.
12 Openshaw, "Touchy Subject," 161.
13 Eben Bayer quoted in Openshaw, "Touchy Subject," 161.
14 Koert van Mensvoort and Hendrik-Jan Grievink, *Next Nature: Nature Changes Along With Us* (Barcelona: Actar, 2011), 460–61.
15 National Nuclear Security Administration, "Program Statement of the Advanced Simulation and Computing (ASC) Predictive Science Academic Alliance Program (PSAAP)," accessed January 24, 2016, http://nnsa.energy.gov/sites/default/files/nnsa/inlinefiles/program_statement.pdf.
16 Kevin Kelly, *What Technology Wants* (New York: Viking, 2010), 273.
17 Steven Johnson, *Where Good Ideas Come From: The Natural History of Innovation* (New York: Riverhead Books, 2010), 19.
18 Clayton M. Christensen, Scott D. Anthony, and Erik A. Roth, *Seeing What's Next: Using the Theories of Innovation to Predict Industry Change* (Cambridge, MA: Harvard Business Review Press, 2004), 5.
19 Schumpeter, "Disrupting Mr. Disrupter," *Economist*, November 28, 2015, http://www.economist.com/news/business/21679179-clay-christensen-should-not-be-given-last-word-disruptive-innovation-disrupting-mr.
20 Matt Rogers, "Energy = innovation: 10 disruptive technologies," *McKinsey on Sustainability & Resource Productivity*, no. 1 (Summer 2012): 14.
21 Moskowitz, *Advanced Materials Revolution*, 15.
22 Ibid., 18.
23 See accessed January 24, 2016, http://www.csinet.org/masterformat for more information.
24 Maria Popova, "The Usefulness of Useless Knowledge," *Brain Pickings*, accessed January 24, 2016, http://www.brainpickings.org/index.php/2012/07/27/the-usefulness-of-useless-knowledge/.

1

Concrete

> Concrete Remixed

> Concrete is the most heavily used
> material on earth after water,
> and yet it is also the most
> imperiled construction material.
> Roughly twenty-five billion tons of
> concrete are produced annually —
> more than twice the amount of all
> other building materials combined —
> contributing between 5 and 10
> percent of all global CO_2 emissions.[1]
> Despite this significant environ-
> mental footprint, concrete is
> challenging to recycle, and the
> global consumption of cement
> continues to increase between 2 and
> 4 percent annually.[2] Meanwhile,
> steel-reinforced concrete, which
> comprises the bulk of concrete
> construction, is woefully imper-
> manent.

According to Chicago-based Klein and Hoffman Senior Principal and structural engineer Jay Paul, "The extent of deterioration to concrete structures globally is occurring at an alarming rate, which challenges engineers on this continent and throughout the world on a daily basis."[3] Realizations of concrete's environmental impact and premature technical obsolescence have motivated scientists, manufacturers, and the building community to rethink this ubiquitous substance. As a result, next-generation concrete offers multiple approaches to solving these problems and contributes compelling new ideas for concrete applications.

Concrete's popularity began with the realization that humans could manipulate a fluid, plastic material that would quickly assume a stone-like permanence. Roman engineers made famous use of hydraulic concrete to create a wealth of ambitious civic and infrastructural projects before the empire's fall in the fifth century. Remarkably, we still use many of these structures today: buildings, bridges, roads, and aqueducts that have survived roughly two millennia of wear and tear. The Pantheon, for example, which exhibited the longest structural span for over seventeen centuries, today remains a popular tourist destination with only minimal cracking evident in its dome.[4]

After Rome's decline, concrete fell into disuse for centuries, until the cultivation of portland cement and reinforced concrete resulted in the material's resurrection by the mid-nineteenth century.[5] Since then, so-called liquid stone has come to represent—both conceptually and literally—the making of our modern physical environment. As historian Antoine Picon states, "No material has been more closely associated with the origins and development of modern architecture than concrete."[6] For all its advantages—strength, manipulability, and practically omnipresent ingredients—contemporary reinforced concrete proliferated too quickly without a serious assessment of its drawbacks. Specifically, portland cement and steel reinforcing are problematic elements from ecological and durability standpoints, respectively.

Portland cement is responsible for 94 percent of concrete's CO_2 emissions but occupies only 10 to 20 percent of its volume.[7] Humanity's insatiable demand for concrete, which continues unabated, caused cement plant emissions to nearly triple between 1990 and 2006 due to increased production, with an estimated CO_2 contribution of 2.34 billion tons by 2050.[8] As a result, the concrete industry is actively seeking ways to reduce the material's hefty ecological footprint. **↙ One method entails replacing some or all of the portland cement with alternative cementitious materials, such as fly ash or slag, which are byproducts of coal combustion and metal production, respectively (see page 46).** According to one study, slag cement mixtures can reduce CO_2 emissions by up to 46 percent.[9] Another strategy involves capturing industrially emitted CO_2 for use in manufacturing calcium carbonate-based cement—effectively sequestering the greenhouse gas.[10] Improvements may also be made at the other end of the life cycle, with enhancements to concrete recycling (see page 42).

Steel inherently oxidizes over time, due to the extremely reactive nature of iron, which comprises 88 to 98 percent of steel. Thus, steel-reinforced concrete is an intrinsically temporary solution, as the steel eventually comes into contact with water, air, and chemicals once its protective concrete covering begins to wear. Because steel oxidation often occurs unnoticed, this condition can lead to surprising and calamitous results. One preventative measure involves the use of alternative reinforcing materials that delay or eliminate the process of ferrous corrosion, such as galvanized steel, epoxy-coated steel, stainless steel, or glass fiber-reinforced polymers. Norway-based ReforceTech manufactures reinforcing bars made of continuous basalt fiber (CBF), a noncorroding material formed into helical shapes with a polymer resin finish.[11] Singapore-based Future Cities Laboratory is experimenting with reinforcing made from woven-strand bamboo (WSB), consisting of decay-resistant carbonized bamboo coated in a water-based adhesive.[12] **↘ Another response is to prolong concrete's useful life, either by maintaining existing concrete structures or creating novel iterations with self-healing properties (see page 28).** Several self-maintaining concrete technologies are currently being tested in real-world applications. These include shape-memory polymers, inorganic healing agents, and capsules containing bacterial healing agents. According to Cardiff University engineering professor Bob Lark, "These self-healing materials and intelligent structures will significantly enhance durability, improve safety, and reduce the extremely high maintenance costs that are spent each year."[13]

In addition to improving environmental and mechanical performance, designers of next-generation concrete seek to overcome the material's negative reputation as a drab, lifeless substance by exploring enhanced functionality and aesthetic potential. For example, **➔ 12 Blocks is a collection of concrete masonry units designed to provide hospitable microenvironments for habitation by small plants and animals.[14] ➔ Biological Concrete panels offer a similarly biocompatible surface conducive to the growth of lichens and mosses (see page 32).** Additional enhancements include concrete with sun-activated photocatalytic cement that reduces local air pollution, or BlingCrete, a textile-reinforced concrete (TRC) with retroreflective glass microspheres that enhance nocturnal vision (see page 34).

Concrete's ubiquity is both an asset and a pitfall. Because so many concrete structures already exist, we can reuse them adaptively without incurring the environmental costs of producing new concrete. Yet this material foundation is inherently flawed. As *Concrete Planet* author Robert Courland states, "Virtually all the concrete

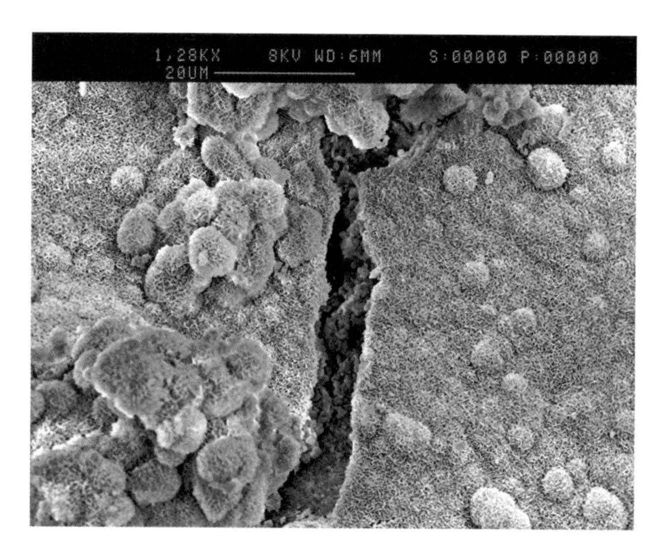

structures one sees today will eventually need to be replaced, costing us trillions of dollars . . . in the process."[15] As a result, the next concrete technologies represent clear departures from the material we have come to know so well. With a reduced carbon footprint, enhanced longevity, and augmented functionality, next-generation concrete bears little resemblance to today's material.

---

1 National Science Foundation, The Cement Sustainability Initiative, *Recycling Concrete* (Conches-Geneva, Switzerland: World Business Council for Sustainable Development, 2009), 3; "How Solid Is Concrete's Carbon Footprint?" May 24, 2009, http://www.sciencedaily.com/releases/2009/05/090518121000.htm.

2 Portland Cement Association, "Global Cement Consumption on the Rise," June 3, 2015, http://www.cement.org/newsroom/2015/06/03/global-cement-consumption-on-the-rise.

3 Jay H. Paul, "Repair, Renovation and Strengthening of Concrete Structures" (white paper based on a lecture), Evaluation and Rehabilitation of Concrete Structures, Mexico City, September 11–13, 2002.

4 Bill Addis, *Building: 3,000 Years of Design, Engineering and Construction* (New York: Phaidon Press, 2007), 51.

5 Ibid., 346.

6 Antoine Picon, "Architecture and Technology: Two Centuries of Creative Tension," in *Liquid Stone: New Architecture in Concrete*, ed. Jean-Louis Cohen and G. Martin Moeller Jr. (New York: Princeton Architectural Press, 2006), 8.

7 Jan R. Prusinski et al., "Life Cycle Inventory of Slag Cement Concrete," 10, paper presented at the CANMET/ACI Eighth International Conference on Fly Ash, Silica Fume, Slag and Natural Pozzolans in Concrete, Las Vegas, NV, May 2004.

8 Emad Benhelal et al., "Global strategies and potentials to curb CO_2 emissions in cement industry," *Journal of Cleaner Production* 51 (July 15, 2013): 145.

9 Prusinski et al., "Life Cycle Inventory," 13.

10 More about Calera Corporation's novel process may be found here: accessed January 15, 2016, http://www.calera.com/beneficial-reuse-of-co2/process.html.

11 See accessed January 15, 2016, http://reforcetech.com.

12 See accessed January 15, 2016, http://www.fcl.ethz.ch/project/bamboo/.

13 "UK's first trial of self-healing concrete," *Cardiff University News*, October 28, 2015, http://www.cardiff.ac.uk/news/view/152733-uks-first-trial-of-self-healing-concrete.

14 Blaine Brownell, *Transmaterial 3: A Catalog of Materials that Redefine Our Physical Environment* (New York: Princeton Architectural Press, 2010), 16.

15 Robert Courland, *Concrete Planet: The Strange and Fascinating Story of the World's Most Common Man-Made Material* (Amherst, NY: Prometheus Books, 2011), 135.

HYBRID CONCRETE CASTING SYSTEM

Augmented Skin

A
cement 600
plaster 600
water 600
(at once)

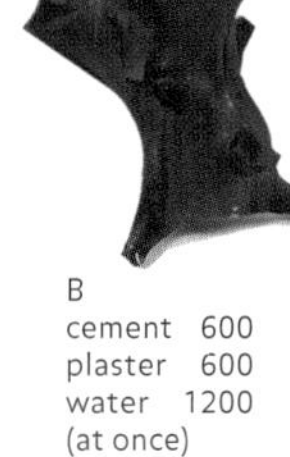

B
cement 600
plaster 600
water 1200
(at once)

C
cement 300
plaster 300
water 600
(every 3 hrs.)

D
cement 150
plaster 150
water 300
(every 30 min.)

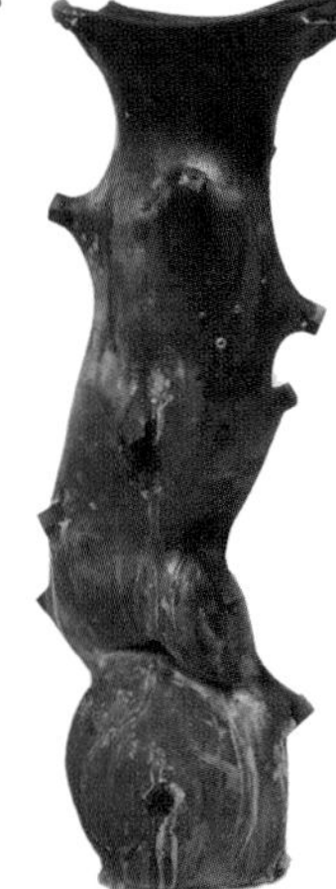

E
cement 450
plaster 450
water 900
(every 45 min.)

Augmented Skin is a custom casting process that enables the creation of complex objects and building components. The technique utilizes different materials in combination to create biomorphic objects that conjoin traditionally disparate elements of structure and cladding. Developed at London's Bartlett School of Architecture by designers Kazushi Miyamoto, Youngseok Doo, and Theodora Maria Moudatsou, the project incorporates two design logics: strand and skin. The strand is based on the human body's skeleton; it is a string-like form that functions as the primary framework. The skin serves as the object's surface and is situated between strand components.

Before fabrication, digital simulation is utilized for the surface deformation design, the distribution of the strand components, and structural testing. Internal cross-shaped strands function as the structural reinforcing for the casting material. These are covered in an elastic fabric that is coated with polyvinyl alcohol (PVA), and the resulting semirigid, self-supporting tension system serves as the formwork for concrete casting. Once the concrete is cast and cured, the final object bears an uncanny resemblance to vertebrate organisms whose structure and skin reveal a much more intimate relationship than in conventional buildings or furniture.

CONTENTS
Concrete, wood framing, fabric

APPLICATIONS
Building components, furniture, sculptural objects

TYPES / SIZES
Various casting materials possible, including cement and plaster

LIMITATIONS
Has not been tested at a building scale

FUTURE IMPACT
Increased integration of structure and skin in buildings and assemblies, emulation of biological systems and processes in building construction

COMMERCIAL READINESS
● ○ ○ ○ ○

CONTACT
The Bartlett
UCL Faculty of the Built Environment
Gower Street, London WC1E 6BT, UK
+44 (0) 2076792000
www.bartlett.ucl.ac.uk
m.dhesi@ucl.ac.uk

CONCRETE-HEALING MICROBIAL PATCH

BacillaFilla

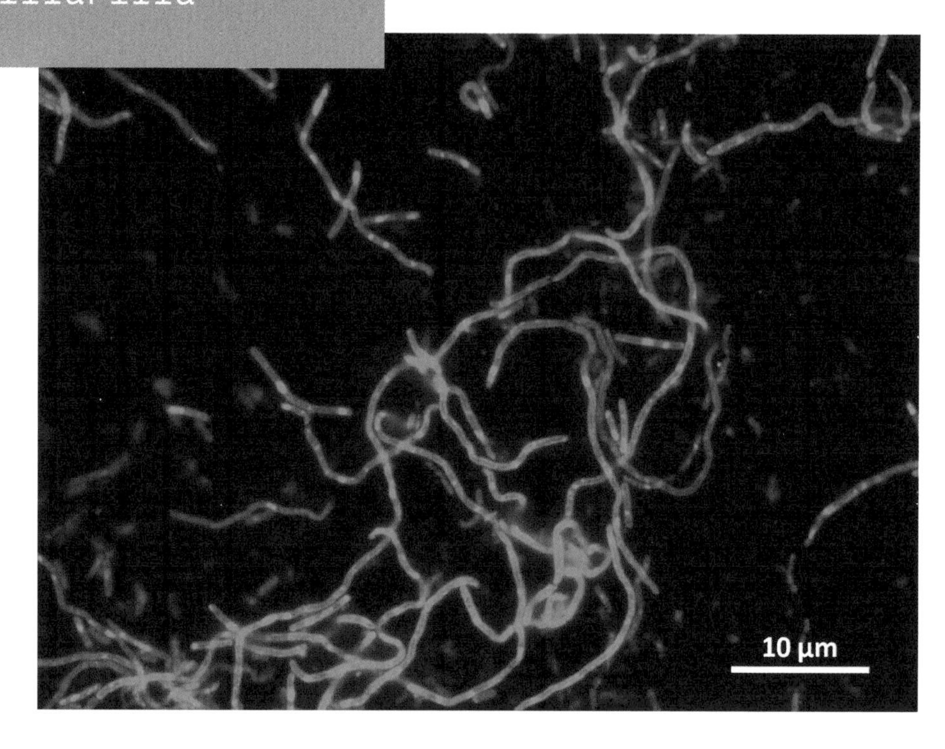

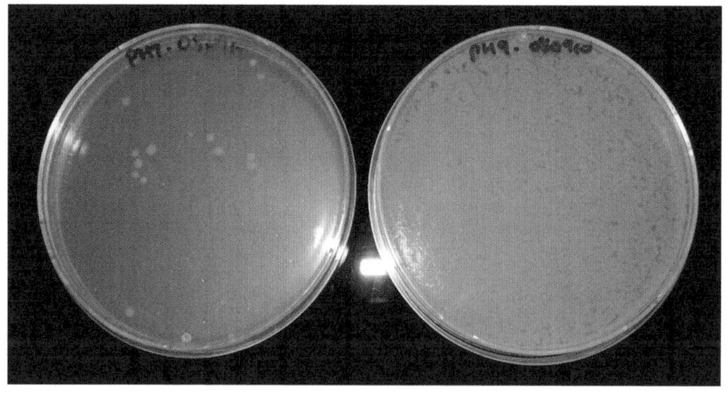

BacillaFilla is an engineered microbial glue developed to repair cracks in concrete that can cause catastrophic structural failure. The result of a multidisciplinary research effort at Newcastle University, the material is designed to be spray applied near tectonic fissures. Once on the concrete surface, the spores begin to germinate and the cells swarm into the crack. When the cell density increases within the fissure, the cells activate concrete repair in three ways. They form crystals of calcium carbonate, reinforcing in the form of fibrous cells, and a natural binding agent called levan adhesive.

Grown in a batch bioreactor, BacillaFilla is created when cells undergo sporulation, and the spores are then transferred to storage containers. Because the spores do not require nutrition or constant care, they are ideally suited for long-term storage and transportation. Once sprayed on site, the microbial patch is designed to cure at roughly the same strength and pH as conventional concrete and includes a kill switch so the germination process may be terminated.

CONTENTS
Bacillus subtilis 168, 3,610 engineered microbes

APPLICATIONS
Concrete repair

TYPES / SIZES
Vary

ENVIRONMENTAL
Prolong functional life of concrete structures

LIMITATIONS
Not suitable for large cracks, high-strength concrete, or fair-faced concrete

FUTURE IMPACT
Marriage of the fields of genetic engineering and construction, increased use of living organisms in place of inert materials

COMMERCIAL READINESS
● ○ ○ ○ ○

CONTACT
Newcastle University Center for Bacterial Cell Biology
Richardson Road, Newcastle upon Tyne NE17RU, UK
+44 (0) 1912083203
www.ncl.ac.uk/cbcb/

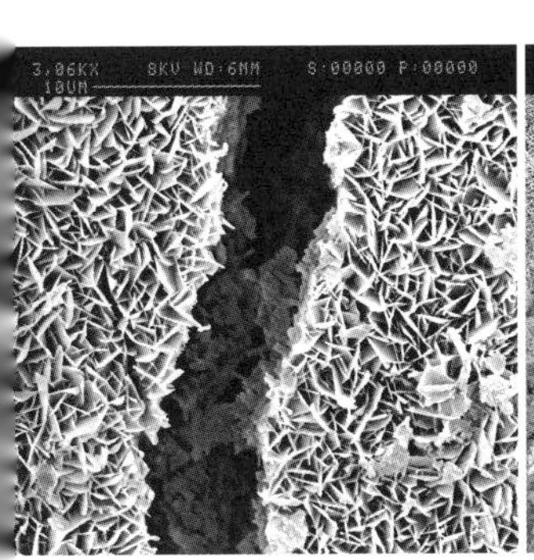

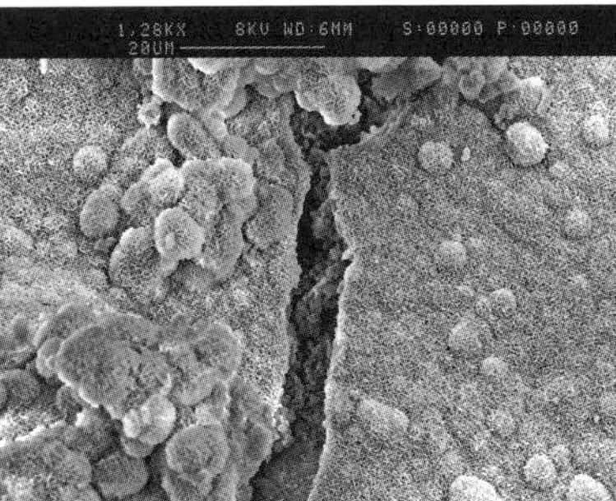

030130–001

ULTRADUCTILE ENGINEERED CEMENTITIOUS COMPOSITE

Bendable Concrete

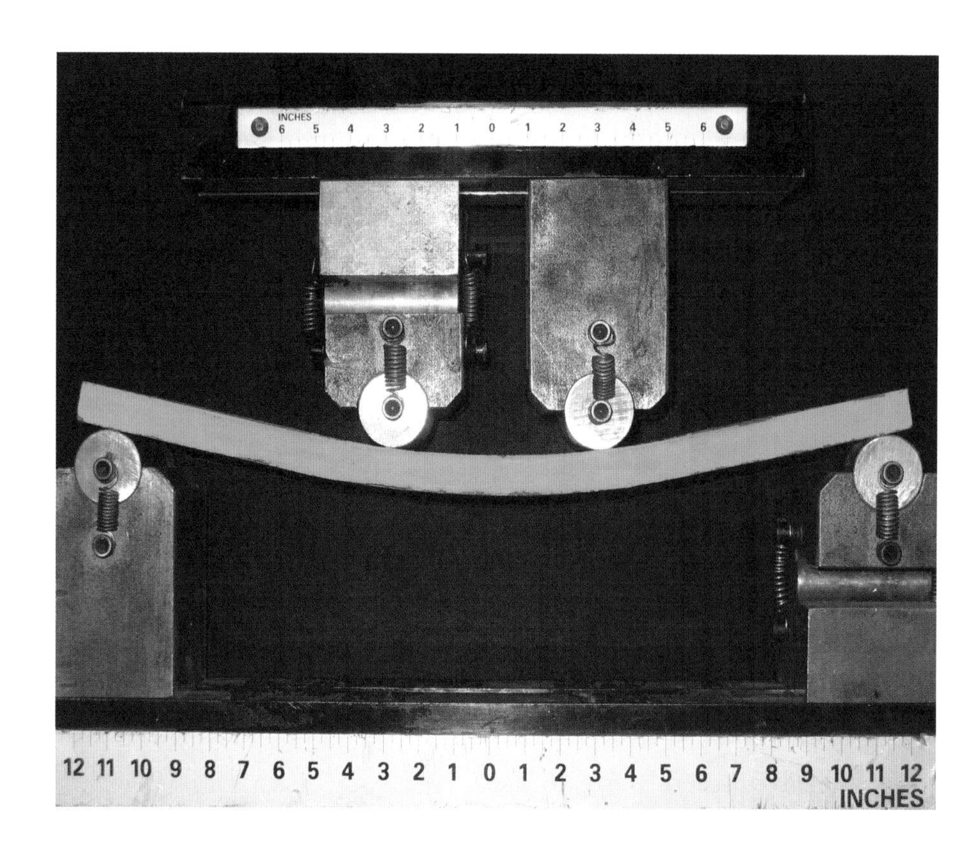

Researchers at the University of Michigan Advanced Civil Engineering Materials Research Lab have developed an engineered cementitious composite (ECC) that is far less brittle than conventional concrete. Made with thousands of short, randomly placed coated microfibers, the ECC can withstand four times the tensile force of its standard counterpart. It also exhibits three hundred times the ductility of regular concrete, hence its popular nickname, bendable concrete.

Under sufficiently high tensile stress, ECC does experience cracking, but the cracks are microscaled—approximately the size of a human hair—and able to self-heal in the presence of air and water. The material also gives way to corroding and expanding reinforcing steel, a phenomenon that commonly causes failures in conventional reinforced concrete. Furthermore, ECC has self-sensing capabilities that allow it to report increased strain and developing structural problems.

Test ECC applications have been successfully installed on highways and high-rise buildings, and the Michigan researchers speculate that a roadbed constructed with ECC could expect a service life of one hundred years—four to five times the life of standard concrete pavement.

CONTENTS
Cementitious composite with short microfibers

APPLICATIONS
Roadways, bridges, high-rise towers, and other structural applications

TYPES / SIZES
Vary

ENVIRONMENTAL
Increased longevity of resources, reduced maintenance; reduces environmental impact by replacing conventional expansion joints

TESTS / EXAMINATIONS
In situ tests conducted on highway M-14 in Michigan and in the Glorio Roppongi apartment tower, Tokyo

FUTURE IMPACT
Widespread reduction of failures in concrete-based infrastructure and building construction, increased life cycle performance and reduced carbon footprint for concrete

COMMERCIAL READINESS

CONTACT
University of Michigan
Advanced Civil Engineering Materials Research Lab
2350 Hayward Street, Ann Arbor, MI 48109
734-647-7000
www.umich.edu/~acemrl
engin-info@umich.edu

BIORECEPTIVE CONCRETE PANELS

Biological Concrete

The Structural Technology Group at the Universitat Politècnica de Catalunya (UPC) has developed a new type of concrete that supports the growth of small mosses, fungi, lichens, and microalgae. The multi-layered material is composed of two types of cement: standard portland cement and magnesium phosphate cement, which is slightly more acidic and thus conducive to biological growth. These materials are encapsulated by additional layers to make a vertically oriented panel for building facades: a waterproofing layer that protects the cementitious materials from damage, a biological layer that absorbs water and supports plant colonization and growth, and a reverse waterproofing coating that facilitates the collection and storage of rainwater in the organic stratum.

The aim of biological concrete is to support natural, comprehensive colonization within a year. The growth of pigmented organisms that change their color and texture over time brings a fresh, transformative character to conventionally homogeneous and static building facades.

CONTENTS

Carbonated concrete, magnesium phosphate cement, waterproofing, biological support layer, discontinuous coating

APPLICATIONS

Building envelopes, living walls, decorative surfaces

TYPES / SIZES

Vary

ENVIRONMENTAL

Supports growth of lichens, mosses, fungi, and microalgae; reduces atmospheric CO_2 via biological coating; high thermal mass

TESTS / EXAMINATIONS

Experiments conducted at the UPC and Ghent University in Belgium

FUTURE IMPACT

Enhanced biological support in buildings beyond green roofs and vertical gardens

COMMERCIAL READINESS

CONTACT

Universitat Politècnica de Catalunya
Structural Technology Group
Jordi Girona, 1-3, Campus Nord, Edif. C-1,
Barcelona 08034, Spain
+34 934017351
https://dec.upc.edu
info.ec@upc.edu

RETRORELFECTIVE CONCRETE

BlingCrete

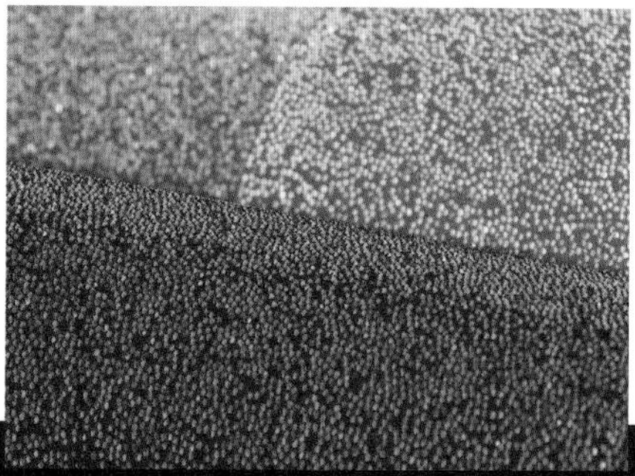

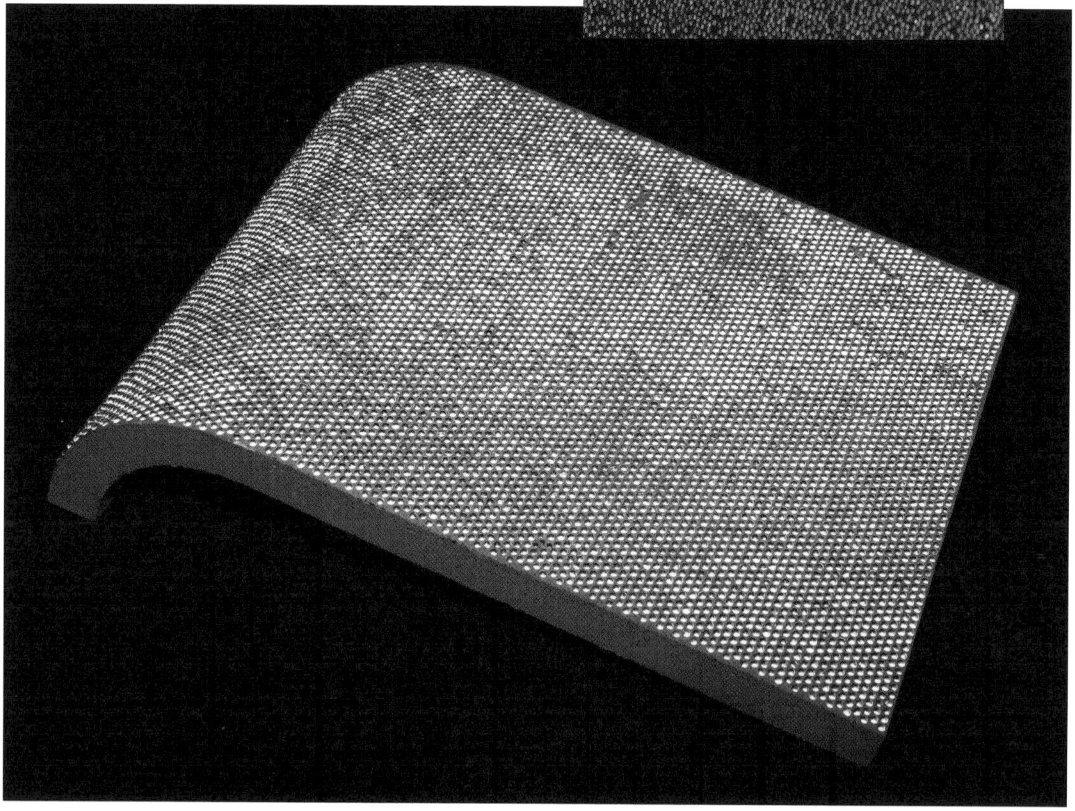

PRODUCT

BlingCrete is a textile-reinforced concrete designed to reflect light back toward its origin. The precast panels are fabricated with an outer layer of glass microspheres that enable retroreflectivity, an optical phenomenon in which light is reflected back toward the direction from which it appears.

Many other retroreflective materials are exclusively fabric or film based; thus, they are not as durable, permanent, or inflammable as textile-reinforced precast concrete. Also, these materials are typically applied in narrow swatches, as opposed to the large, continuous surface provided by BlingCrete.

The concrete's reflective properties open up various design possibilities in architecture, interior design, and infrastructure. Potential applications include safety-related marking of danger spots in construction areas, integrated guidance systems, and cutting-edge surface components.

CONTENTS

Textile-reinforced concrete, glass microspheres

APPLICATIONS

Architectural cladding, pavement, urban furniture, signage, integrated guidance systems

TYPES / SIZES

Standard, high- or ultraperforming concrete; grayscale variations from white to anthracite, custom colors and patterns available; glass bead sizes $^1/_{32}$–$^1/_4$" (0.7–7 mm) diameter

ENVIRONMENTAL

Passively increases intensity and range of light sources

TESTS / EXAMINATIONS

Stability, skid resistance, grip / SRT pendulum, fire, alkali resistance, and antigraffiti tests

FUTURE IMPACT

Multifunctional concrete surfaces with new properties; decreased electricity consumption due to retroreflective materials

COMMERCIAL READINESS

CONTACT

BlingCrete
Rochstrasse 9, Berlin 10178, Germany
+49 1728110266
www.blingcrete.com
info@blingcrete.com

AIR-PURIFYING CONCRETE MASONRY SYSTEM

Breathe Brick

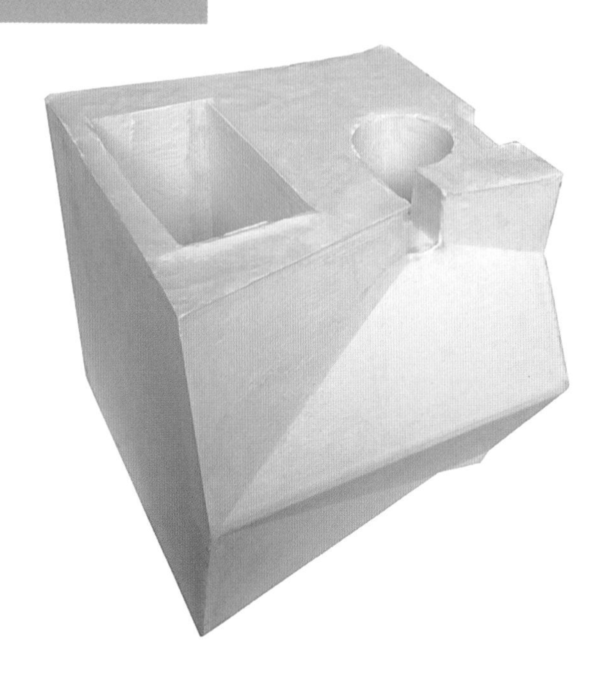

PRODUCT

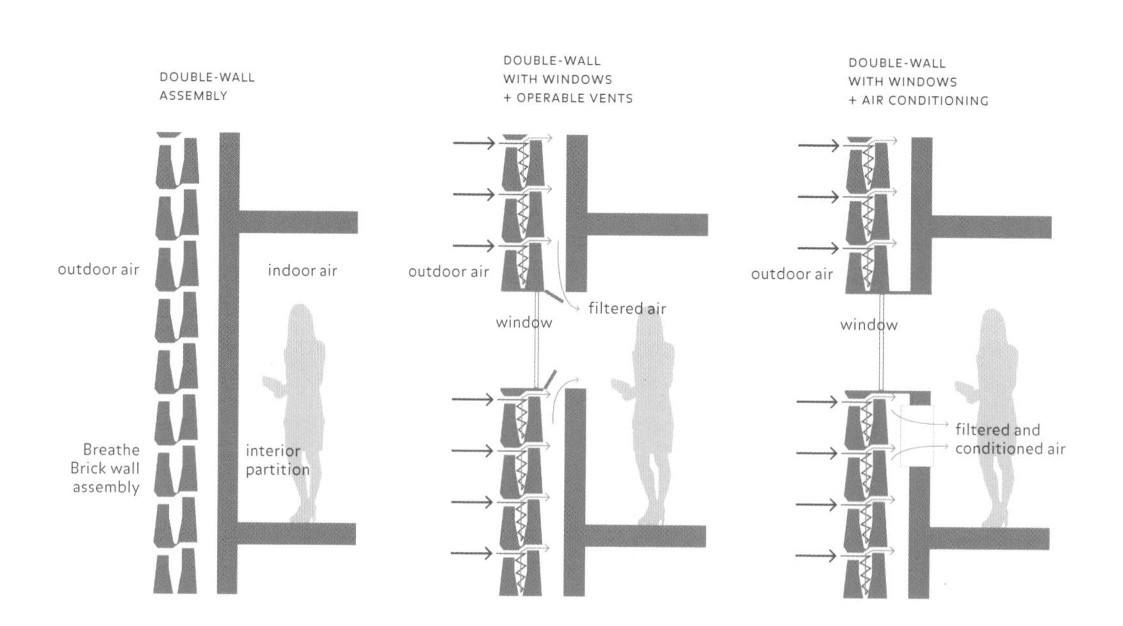

Breathe Brick is a masonry system that filters polluted outdoor air so that it becomes healthy enough to bring directly into occupied spaces. Based on the principle of cyclonic technology, in which physical particles are removed from a fluid via vortex separation, Breathe Brick does not require any energy to operate, as it works with pressure and temperature differentials. Air flows into the individual modules by way of faceted surfaces, and an embedded cyclone filter causes air to create a small vortex, filtering out impurities in the process. The modules are vertically aligned via recycled polymer couplers, and particulates fall into a collection hopper in the bottom course.

Developed by California Polytechnic State University architecture professor Carmen Trudell, the system is a passive method of improving indoor air quality as well as an elegant load-bearing exterior wall assembly that can be customized or mass-produced based on the application. Breathe Brick is optimal for one- or two-story structures in areas with poor air quality.

CONTENTS

Porous concrete, cyclone filter, recycled plastic coupler

APPLICATIONS

Load-bearing or freestanding masonry walls

TYPES / SIZES

Replaces standard 8" concrete masonry unit

ENVIRONMENTAL

Passive filtering of air pollution

LIMITATIONS

Does not provide a thermal or moisture barrier, so must be used with a double-wall assembly

FUTURE IMPACT

Net-zero air pollution filtering at the building envelope, improved indoor air quality standards—especially in polluted and economically disadvantaged areas

COMMERCIAL READINESS

● ○ ○ ○ ○

CONTACT

California Polytechnic State University
1 Grand Avenue, San Luis Obispo, CA 93407
805-756-1316
www.architecture.calpoly.edu
architecture@calpoly.edu

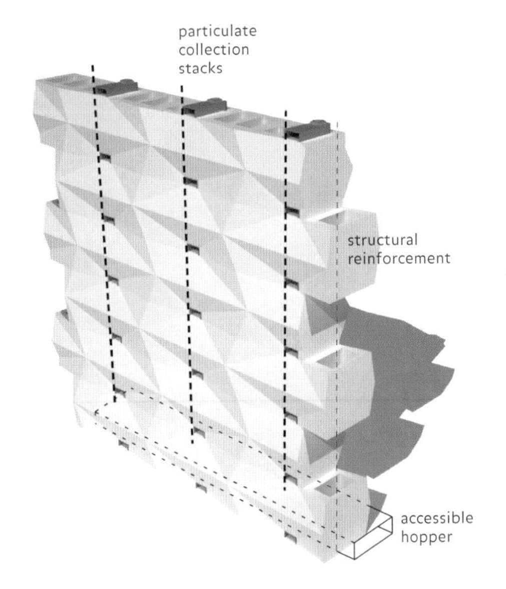

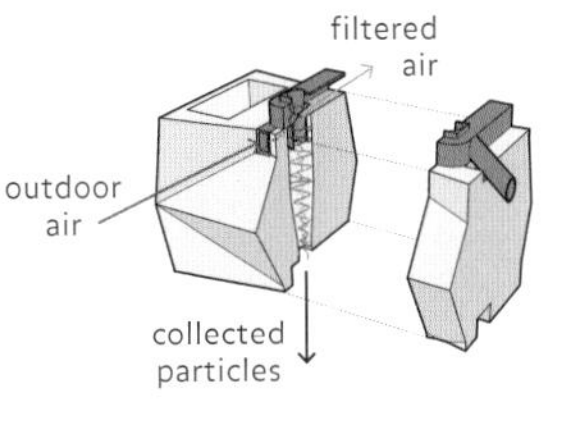

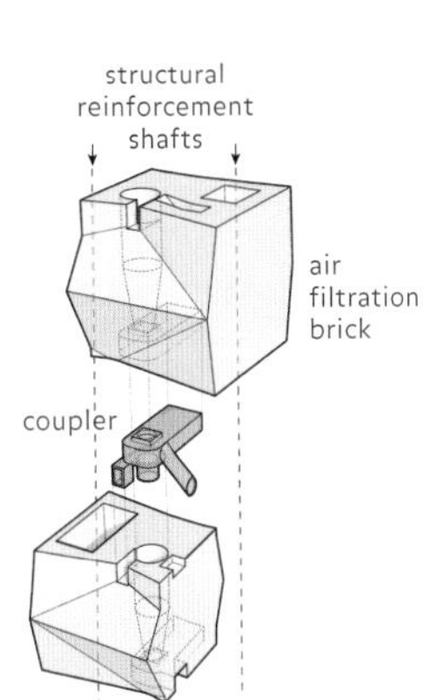

INTELLIGENT

ENERGY-GENERATING CONCRETE

DysCrete

DysCrete is a dye-sensitized energy-generating concrete made by coating pre-fabricated concrete components with layers of organic dyes held in suspension. Dye-sensitized solar cells (DYSCs) rely on a photoelectrochemical process to convert light into energy. Made of light-reactive natural dyes from substances like spinach chlorophyll, the cells generate power that is harnessed by oxide electrodes. This technology has the advantages of low cost and flexible applicability, but its long-term stability is a challenge to optimize.

In DysCrete, the DYSC cell is applied via a hybrid process of sintering and spray deposition in multiple stages. This layered approach results in a more reliable system than the conventional DYSC encapsulation process, in which any imperfection will render a cell useless. The system may be tuned to specific wavelengths of light by manipulating the dye and electrolyte components, and can generate electricity from indirect as well as direct light sources.

CONTENTS

Concrete, organic dyes, electrolytes

APPLICATIONS

Precast concrete blocks and panels for building envelopes

TYPES / SIZES

Vary

ENVIRONMENTAL

Energy-harvesting capabilities in varied light conditions

TESTS / EXAMINATIONS

Laboratory tests, exhibited at BAU 2015

LIMITATIONS

Long-term instability of the DYSC cell

FUTURE IMPACT

Giving the world's most widely used building material the ability to generate renewable energy

COMMERCIAL READINESS

● ○ ○ ○ ○

CONTACT

BlingCrete
Rochstrasse 9, Berlin 10178, Germany
+49 1728110266
www.blingcrete.com
info@blingcrete.com

CONCRETE WITH SPATIALLY VARYING MICROSTRUCTURE

Functionally Graded Printing

PROCESS

Functionally graded materials, or materials with spatially varying composition or microstructure, are ubiquitous in nature. Examples include human bones, palm trees, or bird beaks—all of which exhibit optimal structural efficiency. In human-made structures such as concrete columns, however, materials are commonly homogeneous. Although physical uniformity enables easier manufacturing and construction of building materials, functionally graded materials demonstrate superior strength-to-weight ratios and use-based material properties.

With this in mind, MIT Media Lab researchers are developing a 3D printer capable of dynamically combining component materials. Working with concrete and UV-curable polymers, the scientists aim to create structures, such as a bone-inspired beam, based on the use of functionally graded materials.

CONTENTS
Concrete, UV-curable polymers

APPLICATIONS
Structural framing, lightweight composites

TYPES / SIZES
Vary

ENVIRONMENTAL
Efficient use of materials, high strength-to-weight ratio

TESTS / EXAMINATIONS
Laboratory tests

FUTURE IMPACT
Engineered materials that perform like bone and other biological structures

COMMERCIAL READINESS
● ○ ○ ○ ○

CONTACT
Mediated Matter Group
Massachusetts Institute of Technology Media Lab
75 Amherst Street, Room E14-433B, Cambridge, MA 02139-4307
617-324-3626
matter.media.mit.edu
ked03@media.mit.edu

ELECTRODYNAMIC FRAGMENTATION-BASED CONCRETE RECYCLING

Lightning-Processed Concrete

PROCESS

Although it is possible to recycle concrete, the process typically involves breaking the material down into aggregates that are used as a crude material for roads and other infrastructure. This downcycling prevents an ideal full recovery of the original material for a wide variety of new applications.

Researchers from the Holzkirchen, Germany-based Concrete Technology Group at the Fraunhofer Institute for Building Physics IBP have revived a technique that enables such a recovery. Invented and later abandoned by Russian scientists in the 1940s, the process, called electrodynamic fragmentation, involves subjecting the concrete to artificially triggered lightning bolts. This method allows the material to be separated into its individual components of aggregate and cement. The process works by submerging the material in water and introducing an electrical charge, which seeks the path of least resistance through the material, around component boundaries.

The method currently requires one hour for each ton of recovered concrete, but the scientists estimate a future processing speed of twenty tons per hour. According to lead researcher Volker Thome, lightning-processed concrete will enable an efficient concrete recycling rate of 80 percent, which is ten times the current amount.

CONTENTS
Concrete, electrodes, electrical impulses

APPLICATIONS
Concrete recycling

TYPES / SIZES
Vary

ENVIRONMENTAL
Efficient recovery and reuse of material components from waste concrete

TESTS / EXAMINATIONS
Laboratory tests

LIMITATIONS
Currently limited to processing one ton of concrete per hour

FUTURE IMPACT
Ability to reuse waste concrete completely, resulting in reduced CO_2 emissions compared with manufacturing new material

COMMERCIAL READINESS

CONTACT
Fraunhofer Institute for Building Physics IBP
PO Box 1152, Holzkirchen 83601, Germany
+49 80246430
www.ibp.fraunhofer.de
info@ibp.fraunhofer.de

STRESS-DETECTING CONCRETE

Smart Concrete

PRODUCT

Smart concrete is capable of sensing the increase of stresses and early formation of cracks in concrete. This capacity is achieved by adding short carbon fibers to a standard concrete mix. In the presence of structural flaws or added structural loads, the electrical resistance of the concrete increases. Changes in resistance are measured by electrical probes attached to the concrete. Developed by University at Buffalo scientist Deborah Chung, Smart Concrete is intended to anticipate failures in critical infrastructure such as levees, dams, and tunnels. The material may also be used to detect earthquakes and monitor traffic flow or building occupancy.

CONTENTS
Aggregate, water, cement, carbon fibers, electrical probes

APPLICATIONS
Levees, dams, and other critical infrastructure; building structure and foundations

TYPES / SIZES
Vary

ENVIRONMENTAL
Enhanced longevity for concrete structures

TESTS / EXAMINATIONS
Laboratory tests

LIMITATIONS
Approximately 30 percent more expensive than conventional concrete

FUTURE IMPACT
Increased resilience of buildings and infrastructure via early hazard detection and mitigation

COMMERCIAL READINESS

CONTACT
University at Buffalo Mechanical and Aerospace Engineering
318 Jarvis Hall, Buffalo, NY 14260
716-645-2593
www.mae.buffalo.edu
maechair@buffalo.edu

CEMENT-FREE CONCRETE MASONRY UNIT

Watershed Block

Watershed Block is an ecologically responsive and attractive replacement for the traditional concrete block (CMU), one of the most commonly used building materials. Watershed Materials, a technology start-up in Napa, California, supported by the National Science Foundation, began development of Watershed Block in 2011. The company reduces the cement in structural masonry with an ultra-high-compaction manufacturing process that mimics the way stone is formed in nature, along with novel designs that take advantage of the aluminosilicates found in aggregates around the world.

Watershed Materials initially reduced the cement by 50 percent, but further materials and manufacturing refinements enabled the complete elimination of cement. The manufacturer accomplishes the removal of cement with several mix designs that explore the geopolymerization of natural aluminosilicates and the reuse of industrial waste products to form unique binding combinations.

CONTENTS
85–95 percent locally sourced recycled aggregates; 5–15 percent cement-free binders, including geopolymers and lime / slag / aluminosilicate blend, or 5–10 percent ordinary cement

APPLICATIONS
Weight-bearing structural wall systems (same as traditional concrete block)

TYPES / SIZES
Same as traditional concrete block (CMU):
4 × 8 × 16" (10 × 20 × 41 cm)
6 × 8 × 16" (15 × 20 × 41 cm)
8 × 8 × 16" (20 × 20 × 41 cm)

ENVIRONMENTAL
Cement reductions between 50 percent and 100 percent compared with traditional CMU, reductions in material transportation and water consumption

TESTS / EXAMINATIONS
ASTM C90

LIMITATIONS
Not suitable for areas with high freeze–thaw cycles

FUTURE IMPACT
Reduction or elimination of energy-intensive cement from the ubiquitous concrete block

COMMERCIAL READINESS

CONTACT
Watershed Materials
11 Basalt Road, Napa, CA 94558
707-224-2532
www.watershedmaterials.com
info@watershedmaterials.com

2 Mineral

> Return to Earth

> Minerals, or earth-based materials,
> represent the original substance
> of design. Their use far predates
> recorded history and they carry
> strong associations with the past.
> Earthen materials such as loam, clay,
> and stone are considered ancient
> and therefore primitive, yet science
> has recently produced some of the
> most sophisticated technologies from
> these raw resources. In fact, the
> late twentieth-century emergence of
> so-called advanced ceramics, which
> exhibit more crystallinity, more
> precise microstructure, and superior
> functional characteristics than their
> predecessors, necessitated a new
> material definition by the Versailles
> Project on Advanced Materials and
> Standards (VAMAS).[1]

Yet advanced ceramics and other mineral-based materials have an energy problem. For example, the popular technical ceramic alumina (Al_2O_3), otherwise known as sapphire when in single crystal form, requires up to 27.8 MJ/kg in processing energy—compared to 3.5 MJ/kg primary production energy for brick.[2] For this reason, low-embodied-energy earthen materials are making a comeback, particularly in building construction, with surprising results.

In this chapter, mineral materials include the diverse collection of substances defined as inorganic, nonmetallic solids (apart from concrete and glass, which I discuss in other sections). A significant portion of premodern architecture consisted of rammed earth, stones, bricks, and clay tiles that imparted a gravity and substance to buildings. The properties of these materials began to diverge based on the use of fire: as with glass and concrete, heat processing has provided earthen materials with increasingly desirable characteristics. The application of fire thus marks an important advancement in mineral resources in design and architecture, from earth brick to kiln-fired brick to engineered architectural terracotta. Except for stone masonry, the predisposition toward heat-treated building modules has resulted in the widespread use of "cooked" materials like bricks and tiles for much of the industrialized world's building facades.

From the perspective of technological achievement, intensive processing has produced some remarkable substances. Advanced ceramics and ceramic-based composites are frequently used for demanding applications in the aerospace and automotive industries, as they exhibit ultrahigh strength, high heat tolerance, and unexpected yet desirable qualities such as optical transparency. For example, **↙ transparent armor is an ultrahard transparent ceramic that may be used for bulletproof glass and other applications that require extreme physical robustness (see page 70).** Nanotechnology has also played a critical role in the development of materials with unprecedented attributes. The most recent supermaterial is Aerographene, or graphene-based aerogel, which is the lightest solid ever made—7.5 times lighter than air (see page 54). Another marvel is flexible vanadium pentoxide, a ceramic that may be rolled up or folded without breaking, just like paper (see page 58).

Yet energy-intensive production comes with a cost. According to a 2012 Johns Hopkins University report, the global brick industry produces significant CO_2 emissions, and the black carbon emitted by brick kilns is one of the largest factors in global warming after CO_2.[3] Meanwhile, construction with raw earth—such as rammed or earth brick structures—continues throughout much of the world. According to environmental building consultants Lynne Elizabeth and Cassandra Adams, between one-third and one-half of the global population resides in earth structures.[4] In comparison with engineered brick, nearly all the embodied energy of mud brick is based on human labor. Awareness of this stark difference, in combination with new fabrication technologies, has motivated a renewed enthusiasm for raw earth construction in the developed world.

This shift promises a remarkable new trend: mineral-based additive manufacturing, or geoprinting. Italian manufacturer WASP (World's Advanced Saving Project) has developed a mobile, large-scale digital fabrication strategy to print structures from mud. ↓ **The Big Delta printer is 39' 4" (12 m) tall and can print entire buildings from raw materials on site using mud-extruding technology.**[5] The printer consumes between 1 and 1.5 kW of power, which is little enough to be provided by off-grid solar panels. Thus, the technology is conducive to producing low-cost housing in remote areas, without the need to transport major building materials to the site. A more robust off-site solution is the D-Shape automated process, which prints load-bearing stone structures in one-quarter of the time of conventional construction methods (see page 60).

Architects have developed enterprising strategies to mass-produce mineral structures using sand and bacteria. This process is based on the behavior of *Sporosarcina pasteurii* bacteria, which employ biological cementation to calcify sand into sandstone. Using this method, manufacturer bioMason makes masonry units out of biomanufactured cement (see page 56). The bio soil–generated bricks possess the same strength as conventional masonry and have been successfully produced using seawater. ↗ **Design firm Ordinary Vis has proposed an urban-scale project using this process, in which the Sahara Desert can be seeded with *Sporosarcina pasteurii* to produce a communal settlement called Dune City.**[6] This bioterraforming strategy, while ambitious, demonstrates how microbial builders might transform an expansive landscape.

New mineral technologies also include creative methods for recapturing resources. For example, StoneCycling's WasteBasedBrick is made from materials salvaged from demolished buildings and industrial waste that would otherwise be deemed unrecyclable (see page 72). Studio Roosegaarde practices another form of reclamation, using an air-purifying tower that cleans 30,000 m^3 of air hourly to also sequester air pollution in the form of dense cubes of highly compressed smog particles that are then turned into jewelry (see page 66).

These sophisticated, aspirational approaches to producing and recapturing mineral building products point to a renewed focus on earthen materials. Two trajectories—one generating energy-intensive supermaterials, and one creating low-energy in situ earth architecture—define the next generation of mineral material technologies.[7] Each has advantages and drawbacks: The former has produced genuine scientific marvels, yet is hindered by its carbon footprint. The latter presents a provocative future, envisioning a tangible path to reconnecting the building with its site, yet requires further development to achieve the scale of its ambitions. Shortcomings aside, both approaches promise an imminent change in cultural attitudes toward earthen materials. As architect Ronald Rael states, "Earth buildings are commonly perceived to be used only by the poor or found only in 'developing' countries."[8] With their universal potential, next-generation mineral technologies have the capacity to bring about both technological and cultural transformation.

1 "Developing a unified classification system for advanced ceramics was seen as a critical need in the effort to achieve compatibility among materials databases." See James G. Early and Harry L. Rook, "Versailles Project on Advanced Materials and Standards (VAMAS)," *Advanced Materials* 8, no. 1 (1996): 10.

2 Michael F. Ashby, "Material Profiles," *Materials and the Environment: Eco-Informed Material Choice*, 2nd ed. (Oxford: Butterworth-Heinemann, 2012), 459.

3 Alexander Lopez et al., *Building Materials: Pathways to Efficiency in the South Asia Brickmaking Industry*, Johns Hopkins University, November 2012, 5.

4 Lynne Elizabeth and Cassandra Adams, eds., *Alternative Construction: Contemporary Natural Building Methods* (Hoboken, NJ: John Wiley & Sons, 2000), 88–89.

5 See accessed January 15, 2016, http://www.wasproject.it/w/en/.

6 Blaine Brownell and Marc Swackhamer, *Hypernatural: Architecture's New Relationship with Nature* (New York: Princeton Architectural Press, 2015), 100–101.

7 Obviously, this is a broad generalization made for argumentation purposes. Some methods can be applied to reduce the carbon footprint of ceramic production, such as the use of highly efficient kilns, and a process like 3D printing with stone can be energy intensive.

8 Ronald Rael, *Earth Architecture* (New York: Princeton Architectural Press, 2009), 9.

GRAPHENE-BASED AEROGEL

Aerographene

MATERIAL

Aerogel is a type of material structure created when the liquid in a gel-based substance is replaced with a gas, resulting in an ultralight solid with very low density and thermal conductivity. Invented by chemical engineer Samuel Stephens Kistler in the 1930s, aerogel is known by nick-names such as "frozen smoke" and "solid air," given that it is composed of approxi-mately 98 percent air. The first aerogel was silica based, and until 2011 it was the lightest human-made solid material until lighter carbon-based versions replaced it.

Aerographene is the most significant of these. The current world record holder as the lightest solid ever made, Aero-graphene has an ultralow density of 160 g/m³. For perspective, the density of air at room temperature is 1,200 g/m³, and the density of helium is 179 g/m³; thus, the new material is, remarkably, 7.5 times lighter than air and just lighter than helium—making it the second lightest known substance after hydrogen.

Discovered in 2013 by scientists at Zhejiang University, Aerographene is made using a new method called lyophilization, which involves freeze-drying hybrid solu-tions of carbon nanotubes and graphene, resulting in a carbon sponge. Because this sponge matches the size and shape of its container, Aerographene can be created in sizes up to thousands of cubic centimeters and in a variety of forms.

Initial applications include the cleanup of oil spills, since the material can rapidly absorb up to nine hundred times its weight in petroleum, as well as the manufacture of ultralight composites, insulation, and phase change materials for energy storage.

CONTENTS
Carbon nanotubes, graphene

APPLICATIONS
Environmental remediation, lightweight composites, insulation, energy storage

TYPES / SIZES
Volumes up to thousands of cm³

ENVIRONMENTAL
Ultrahigh material efficiency, proposed use in cleaning oil spills

FUTURE IMPACT
Ultraperforming thermally insulated composites for aerospace, transportation, mobile electronics, and construction applications

COMMERCIAL READINESS
● ○ ○ ○ ○

CONTACT
Zhejiang University
Department of Polymer Science and Engineering
38 Zheda Road, Xihu, Hangzhou, Zhejiang 310027, China
+86 57187951308
www.polymer.zju.edu.cn

BIOLOGICALLY GROWN MASONRY

BioBrick

MATERIAL

The standard architectural material palette consists of many energy-intensive materials, such as brick masonry and concrete. In contrast, nature exhibits notable strategies for creating high-strength materials via biochemical processes that require little energy. North Carolina–based bioMASON has developed a technology based on such an approach, employing bacteria to produce bricks out of sand or other types of aggregate. Similar in composition to sandstone, the resulting BioBricks are manufactured by *Sporosarcina pasteurii* bacteria with the addition of calcium chloride, urea, and yeast extract. These materials are inexpensive, widely available, and ecologically responsible, and they enable the fabrication of bio-based building modules in under five days.

CONTENTS

Calcium chloride, *Sporosarcina pasteurii* bacteria, yeast extract, urea

APPLICATIONS

Load-bearing walls and vaulted ceilings (in lieu of standard brick masonry)

TYPES / SIZES

Standard brick shapes, custom sizes available

ENVIRONMENTAL

Net-zero production of masonry units using abundant, nontoxic materials

TESTS / EXAMINATIONS

Successful tests have been conducted using seawater

LIMITATIONS

Currently limited to small-volume production

FUTURE IMPACT

Dramatic reduction in energy use, fuel consumption, and CO_2 emissions associated with conventional kiln-fired brick; industrial materials will increasingly be grown instead of heat processed

COMMERCIAL READINESS

CONTACT

bioMASON
15 T. W. Alexander Drive, Durham, NC 27703
919-541-9366
www.biomason.com
inquiry@biomason.com

ELASTIC VANADIUM PENTOXIDE PAPER

Ceramic Paper

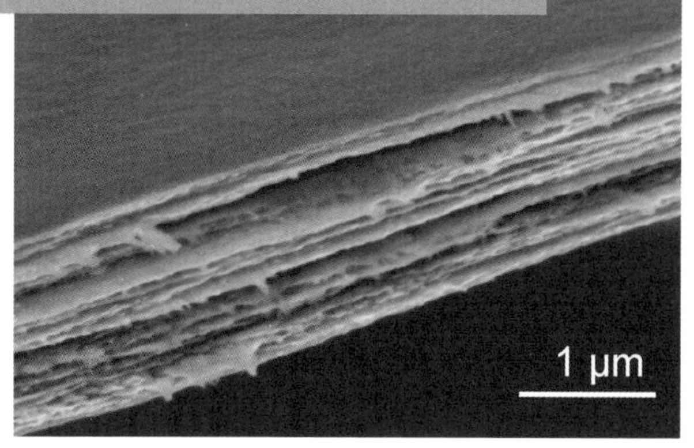

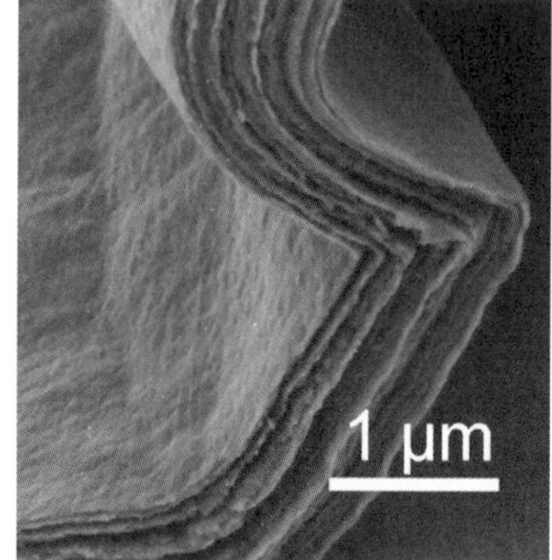

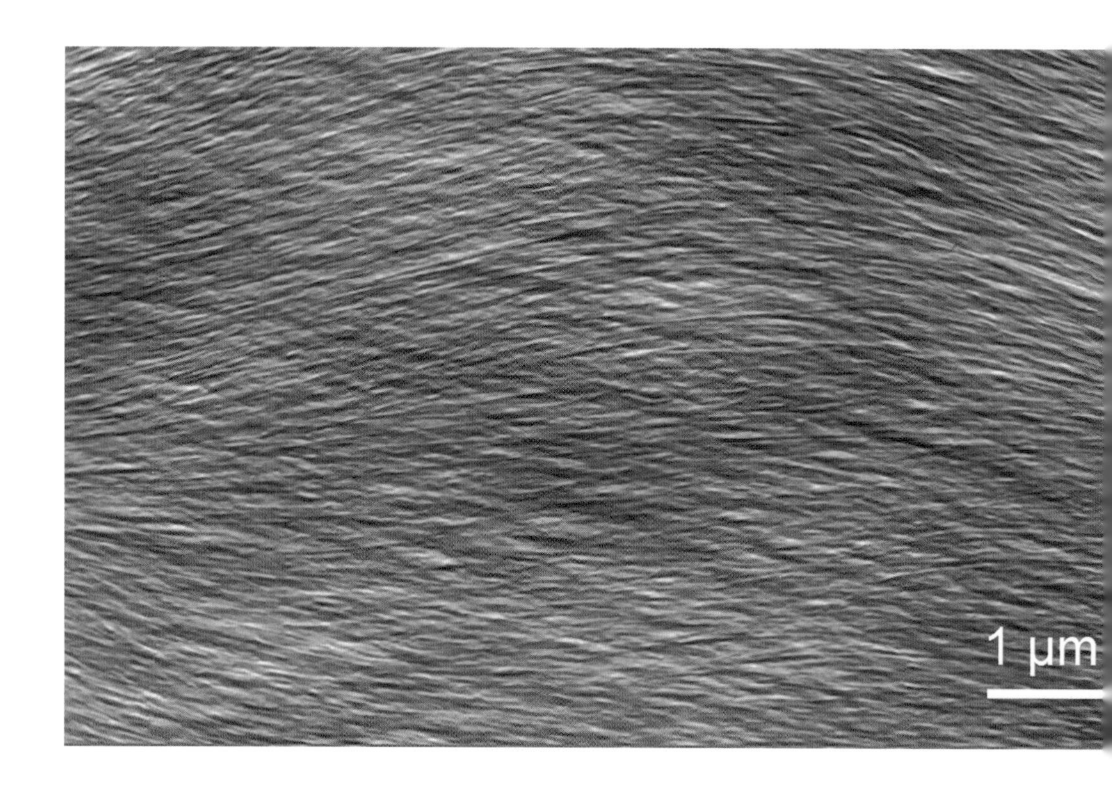

Ceramics are among the oldest materials manipulated by humans, and their properties of rigidity and brittleness are quite familiar. Scientists at the University of Stuttgart Institute for Materials Science have demonstrated that ceramics can be made to behave in unexpected ways—such as to emulate the behavior of paper.

Ceramic paper is flexible enough to be folded or rolled up like paper, yet is stronger than mother-of-pearl. It is also electrically conductive—another characteristic that distinguishes it from conventional ceramics. The researchers make ceramic paper by synthesizing vanadium pentoxide nanofibers in a water-soluble vanadium salt. The fibers aggregate to form a paper-like film when distributed across a substrate and align themselves in parallel orientations to create a linear grain.

The material is not only harder than mother-of-pearl, but also exhibits elastic properties. These characteristics, in combination with electrical conductivity, make it suitable for a variety of future uses in batteries, sensors, and artificial muscle actuators.

CONTENTS
Vanadium pentoxide

APPLICATIONS
Artificial muscles, electrodes for batteries, gas sensors

TYPES / SIZES
Nanofiber solution varies between 0.5 and 2.5 micrometers

TESTS / EXAMINATIONS
Laboratory tests at the Max Planck Institute for Intelligent Systems, Stuttgart, Germany

LIMITATIONS
Electrical conductivity is greater along the grain than along the cross-grain orientation

FUTURE IMPACT
Creation of new uses for ceramic materials requiring elasticity, conductivity, and other unexpected attributes

COMMERCIAL READINESS
● ○ ○ ○ ○

CONTACT
**University of Stuttgart
Institute for Materials Science
Heisenbergstrasse 3, D-70569 Stuttgart, Germany
+49 07116893311
www.uni-stuttgart.de/imw/**

3D PRINTED STONE

D-Shape

PROCESS

D-Shape is a robotic building system that uses sand to print artificial stone structures. Developed by UK-based Monolite, the D-Shape printing technology enables full-size sandstone buildings to be made without human intervention, using a stereolithography 3D printing process that requires only sand and an inorganic binder to operate.

According to Monolite founder Enrico Dini, existing materials such as reinforced concrete and masonry are expensive and inflexible. The construction of complex surfaces requires the provision of resource-inefficient formwork and costly scaffolding. Furthermore, existing techniques require skilled personnel to refer continually to construction documentation at considerable expense.

By contrast, D-Shape takes advantage of newfound CAD/CAM capabilities, printing an entire structure from bottom to top in a single pass. The binder transforms the sand into a mineral with microcrystalline characteristics that is stronger than portland cement–based concrete. Furthermore, Monolite estimates the system to be four times faster than traditional building methods, with a cost that is 30 to 50 percent lower than manual approaches.

CONTENTS
Sand, binder

APPLICATIONS
Load-bearing stone structures

TYPES / SIZES
Vary, printing occurs in 3/16"–5/16"-tall (5–10 mm) sections

ENVIRONMENTAL
Elimination of the need for steel reinforcing, requires no formwork like concrete casting

TESTS / EXAMINATIONS
Traction, compression, and bending tests; full-scale prototypes, such as Radiolaria by Shiro Studio (pictured)

LIMITATIONS
Curing requires twenty-four hours to complete, annual production capacity of a single machine equivalent to twelve two-story buildings

FUTURE IMPACT
Transformation of the building industry via construction automation; ability to reassemble granular materials into compact stone, in highly customizable geometries

COMMERCIAL READINESS

CONTACT
Monolite UK Ltd.
101 Wardour Street, London W1F 0UN, UK
+44 (0) 7983773529
www.d-shape.com
info@d-shape.com

JAMMED PSEUDOSOLID MATERIALS

Morphable Structures

PRODUCT

Jamming is a phenomenon that allows certain complex fluids to behave like solids when their density increases. Granular materials, like sand or coal, are considered complex fluids. They can attain a jammed state with applied pressure, and they become pseudosolid materials with manipulable geometry and rigidity. The principles of jamming are well established in the scientific literature, yet large-scale experiments employing the phenomenon are rare. Researchers at MIT imagine the potential for shape-changing structures and machines using this approach. Their Morphable Structures are jamming prototypes constructed to gain a better understanding of this effect. Composed of elastic membranes and sand, the pseudosolid structures suggest potential applications such as morphable furniture, a floor that dynamically changes its softness to reduce injury to a falling user, a lamp that becomes brighter when touched, or free-form artistic sculpting.

CONTENTS
Sand, elastic membranes

APPLICATIONS
Furniture, surfaces, light sources, and sculptural objects that can shift between a fluid and a pseudosolid state

TYPES / SIZES
Vary

TESTS / EXAMINATIONS
Laboratory tests

LIMITATIONS
Jamming is limited to complex fluid materials such as granular materials, foams, and glasses

FUTURE IMPACT
Increased use of materials with transformable, phase-changing physical properties

COMMERCIAL READINESS
●○○○○

CONTACT
Mediated Matter Group
Massachusetts Institute of Technology
Media Lab
75 Amherst Street, Room E14-433B,
Cambridge, MA 02139-4307
617-324-3626
matter.media.mit.edu
ked03@media.mit.edu

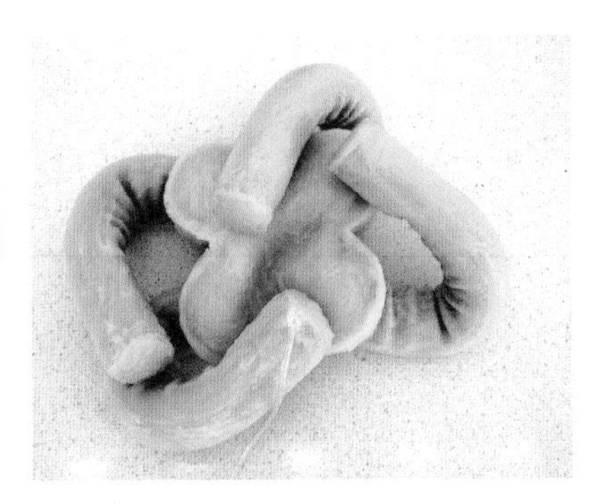

CRACK-REPAIRING BITUMEN

Self-Healing Asphalt

The asphalt binder in a roadbed is continually degraded by environmental forces such as solar ultraviolet rays. In time, the material loses its adhesive ability to keep the surface particles intact. The roadbed eventually deteriorates as moisture permeates the surface, resulting in roughness, potholes, and later structural failure. Typical sealants and rejuvenators can extend the life of a roadbed, but they only operate at the surface and cannot prevent deep decay.

Asphalt exhibits self-healing properties at a high temperature, but only if it is left undisturbed. Researchers at the Delft University of Technology investigated the ability to heat asphalt internally with induction energy to increase its healing rate. The first prerequisite of induction heating is that the heated material must be conductive, which the scientists accomplished by adding electrically conductive fillers and fibers to the bitumen. The second prerequisite is that these fillers and fibers connect in closed-loop circuits. When a microcrack appears in the asphalt, conductive materials form closed-loop circuits around the microcrack. A coil induces eddy currents in the closed-loop circuits, which generate heat and melt the bitumen, thus closing the crack.

CONTENTS
Porous asphalt concrete, steel wool fibers

APPLICATIONS
Self-repairing roadways

TYPES / SIZES
Vary

ENVIRONMENTAL
Extends material service life

TESTS / EXAMINATIONS
Test track on A58 near Vlissingen, Netherlands

FUTURE IMPACT
Self-healing infrastructure, lower-maintenance roadways

COMMERCIAL READINESS

CONTACT
Delft University of Technology
Civil Engineering and Geosciences
Postbus 5, Delft 2600 AA, Netherlands
+31 (0) 152789111
www.tudelft.nl
info@tudelft.nl

INTELLIGENT

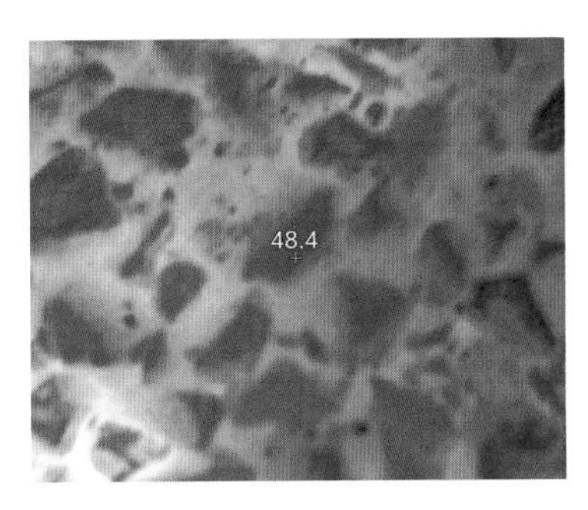

AIR-PURIFYING TOWER THAT CREATES COMPRESSED SMOG

Smog Free Project

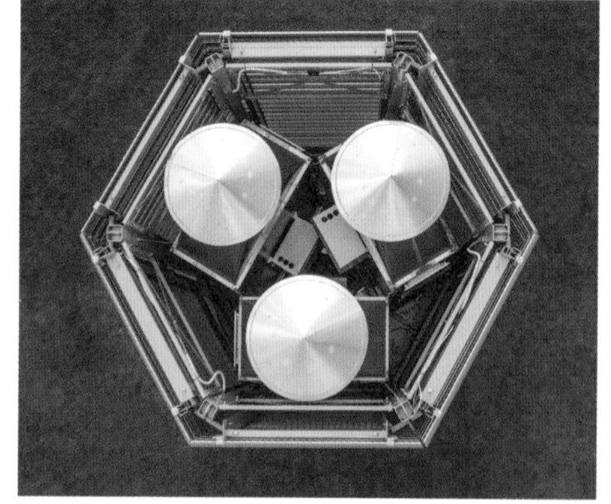

PROCESS

Netherlands-based Studio Roosegaarde has created what it calls the largest vacuum cleaner for smog in the world. Designed as an industrial-strength air purifier for public spaces, the Smog Free Tower employs ion-based filtration technology to create zones of clean air. The fund-raising campaign for the first tower included the sale of rings and cuff links containing dense cubes of compressed smog particles—souvenirs of the eradicated pollution. Consumers who purchased these artifacts indirectly donated 1,000 m³ of clean air to benefit the public.

The Smog Free Project does not represent a final solution; rather, it is an installation that allows visitors to experience the benefits of clean air. In 2014 the first 7 m high Smog Free Tower equipped with low-energy ion technology opened to the public in Rotterdam, and it has since been traveling around the globe. The tower cleans 30,000 m³ of air per hour and runs on 1,400 watts of renewable energy.

CONTENTS
Highly compressed smog particles

APPLICATIONS
Air filtration in parks, gardens, sports facilities, and other sites

TYPES / SIZES
The Smog Free Tower is a 23' × 11' 6" (7 × 3.5 m) modular system; souvenirs include Smog Free Rings, Smog Free Cubes, and Smog Free Cufflinks

ENVIRONMENTAL
Rapid air purification of targeted spaces

LIMITATIONS
Cleans up to 30,000 m³ of air per hour

FUTURE IMPACT
Environmental remediation of public spaces in highly polluted cities

COMMERCIAL READINESS

CONTACT
Studio Roosegaarde
Vierhavensstraat 52–54,
Rotterdam 3029 BG, Netherlands
+31 103070909
www.studioroosegaarde.net
mail@studioroosegaarde.net

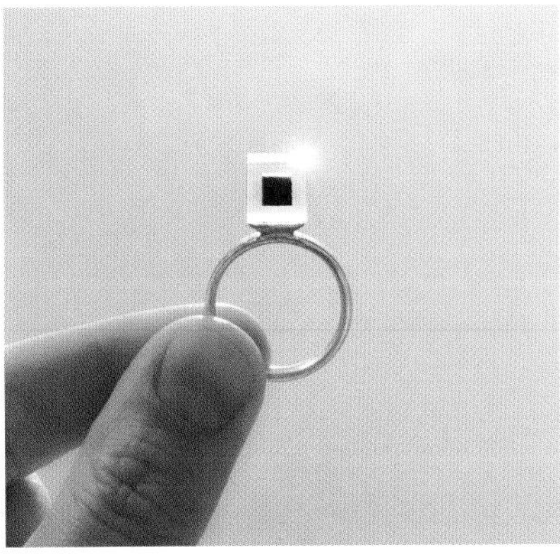

3D PRINTED MINERAL STRUCTURES

Stone Spray

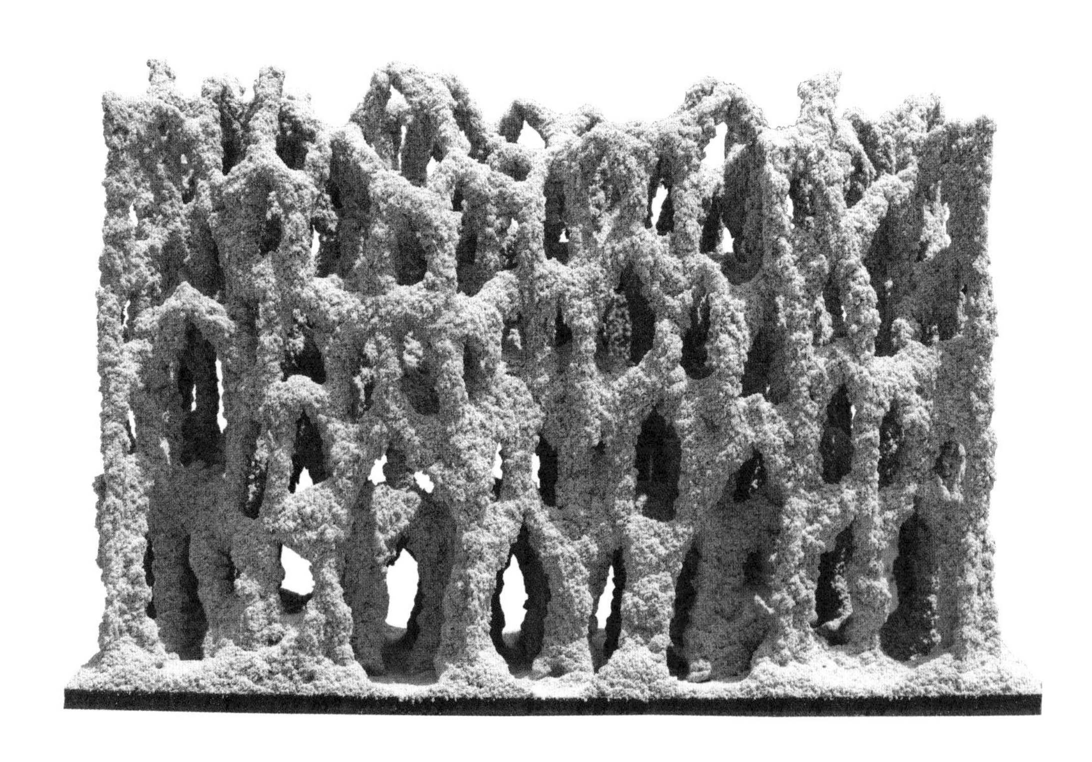

Stone Spray is a building method that creates structures out of soil. Barcelona-based researchers Anna Kulik, Petr Novikov, and Inder Shergill developed the technique while focusing on methods of additive manufacturing in architecture, aiming to create a new environmentally responsible, efficient, and innovative approach to printing structures.

The mechanized Stone Spray printer collects dirt and sand located on site, mixes them with a binder ingredient, and ejects the mixture from a nozzle. The sand-binder mixture solidifies once it makes contact with a surface, such as a scaffold or metal framework. The designer maintains precise control over the resulting shapes based on the printer's digital program. Unlike many 3D printers, Stone Spray prints in multiple directions, including vertically.

CONTENTS
Site-based sand, silt, or fine gravel

APPLICATIONS
Structural fabrication of walls, shelters, bridges, foundations, and furniture; mineral surface applications

TYPES / SIZES
Vary

ENVIRONMENTAL
Uses local materials, low embodied energy

LIMITATIONS
Load-bearing capacity of material determines structural capacity, additional tensile capability requires reinforcing

FUTURE IMPACT
Increased use of on-site materials and renewed interest in earthen construction, hyperlocal building fabrication

COMMERCIAL READINESS

CONTACT
Institute for Advanced Architecture of Catalonia
Pujades 102, Poble Nou, Barcelona 08005, Spain
+34 933209520
www.iaac.net
info@iaac.net

ULTRAHARD SPINEL GLAZING MATERIAL

Transparent Armor

Spinel, or magnesium aluminate, is a durable ceramic material found in some of the world's most renowned gemstones—such as rubies in the British and French Crown Jewels. Spinel occurs naturally in both colored and colorless states, and it can be synthesized in the laboratory as a high-performing replacement for glass, given its significantly higher strength and hardness (its value on the Mohs' scale is 7.5–8.0, compared with 5.5 for glass).

The U.S. Naval Research Laboratory developed a method of manufacturing a polycrystalline version of spinel for transparent armor and bulletproof windows. Unlike glass, wherein a surface crack quickly spreads throughout the material, spinel armor is composed of multiple compressed crystal particles that thwart crack propagation. The NRL fabricates sheets and lens-shaped components via sintering, which involves a low-temperature hot press. The resulting material transmits both visible and infrared light, suggesting suitable uses for imaging systems and windows in cold climates.

CONTENTS

Magnesium aluminate (spinel)

APPLICATIONS

Transparent armor, bulletproof windows, camera lenses, electronic display screens, sensor covers, next-generation lasers

TYPES / SIZES

Various shapes possible; sheets up to 30" (76 cm) maximum dimension

TESTS / EXAMINATIONS

Laboratory tests

LIMITATIONS

Grinding, polishing, and other types of finishing are expensive

FUTURE IMPACT

Highly durable replacement for glass, especially in high-abuse or high-security applications

COMMERCIAL READINESS

CONTACT

U.S. Naval Research Laboratory
4555 Overlook Avenue SW, Washington, DC 20375
202-767-2541
www.nrl.navy.mil
nrl1030@ccs.nrl.navy.mil

BRICKS MADE FROM SALVAGED BUILDING MATERIALS

WasteBasedBrick

WasteBasedBrick originated in Eindhoven, Netherlands, home to the Design Academy, where Tom van Soest was a student. He noted that many buildings in Eindhoven were abandoned and left to deteriorate, the forgotten products of a prior real estate boom.

In 2012 Van Soest founded StoneCycling as a cradle-to-cradle solution for these decaying structures. The company utilizes demolition and industrial waste as raw resources for creating new products with a higher value. StoneCycling collects waste and separates it according to various materials and colors. The company's first offering, WasteBasedBricks, is a collection of bricks available in six different hues, which may be left raw or ground to expose their repurposed mineral composition.

StoneCycling hosts a "Your waste, your brick" program in partnership with real estate owners, municipalities, and architects to develop custom WasteBased-Bricks from existing properties slated for demolition.

CONTENTS
Mineral aggregates from salvaged building materials

APPLICATIONS
Masonry

TYPES / SIZES
Raw (plain) or sliced (ground-faced) versions of six color compositions: Aubergine, Caramel, Mushroom, Nougat, Salami, and Truffle

ENVIRONMENTAL
Repurposing demolition and industrial waste materials into new products, 100 percent recyclable

LIMITATIONS
Color and texture dependent on raw ingredients used

FUTURE IMPACT
Cradle-to-cradle material streams for building construction

COMMERCIAL READINESS

CONTACT
StoneCycling BV
Gedempt Hamerkanaal 111,
Amsterdam 1021 KP, Netherlands
+31 (0) 887774200
www.stonecycling.com
info@stonecycling.com

3 Metal

> From Frame to Filigree

> More than any other material
> category, metal characterizes the
> ages of humanity. The Copper Age
> and Bronze Age (approximately
> 5000 and 1000 BCE, respectively),
> saw the first use of metal tools
> and the dawn of civilization.[1]
> The ensuing Iron Age, roughly
> between 1000 BCE and 0 CE, was
> defined by the pervasive use
> of iron. The much later Steel Age,
> beginning in 1860, marked the
> emergence of the Industrial Revo-
> lution. Thus, our view of tech-
> nological development throughout
> recorded history has been inti-
> mately tied to metal and its
> technical capacities over time.

In architectural history, metal has served as a critical building material, and its role has expanded dramatically over the centuries. Metal was first applied architecturally in two distinct ways: as conspicuous ornament, such as the gold leaf used to gild nails and other objects in ancient Egypt, and as invisible reinforcing, such as small iron cramps used to support stone construction in ancient Greece. These two predispositions for beauty and strength eventually fused in the ferrovitreous architecture of the nineteenth century, in which the technological capability to engineer large-scale iron structures resulted in the first buildings composed predominately of iron and glass. The development of the Bessemer process, which facilitated the mass production of steel, launched the Steel Age and catalyzed the birth of the skyscraper. According to architect Annette LeCuyer, metal became inextricably linked with modernity: "Modernist preoccupations with the skeletal frame and the curtain wall generated the free plan, ideas of universal space, and experiments with mass-produced buildings."[2]

Today metal is entering another period of transition. Steel is one of the most widely utilized materials on the planet, with 2.3×10^9 metric tons produced annually, comprising 12 percent of global energy use.[3] However, change is at hand. The modern epoch marked by the imposing brawn of the steel frame—and the unmitigated consumption of metals in general—is over. A 2015 steel industry report declared that global demand for the ferrous metal has been "evaporating at unprecedented speed."[4] Although this change is partly due to a temporary setback in China's economy, it points to a larger trend with long-term implications. Although they are nonrenewable materials, metals have come to comprise a significant portion of our material palette. Contemporary society's near-total reliance upon nonrenewables has become a dangerously unsustainable position—particularly for rare earth metals —and a gradual shift toward renewable materials is now underway. At the same time, lighter materials are replacing many metals in an effort to reduce energy consumption. For example, carbon fiber composites are gradually supplanting aluminum in vehicles, and the use of structural fiberglass connections is increasing in building facades due to their thermal performance, which is superior to that of metal. Certainly, metal is not disappearing. After all, most of our ninety-two usable elements are metals.[5] Rather, it is experiencing a trend toward dematerialization. The next metal technologies signify a shift from the hefty musculature of the frame to a light and responsive filigree.

Nowhere is this more evident than in the emergence of new supermetals like metallic cellular materials. **← Scientists at HRL Laboratories have made one of the lightest known materials by constructing a microlattice with nickel-phosphorous (see page 94).** The airy substance references large engineering structures, such as the Eiffel Tower—but reduced to a microscale in which individual elements are hollow tubes with only 100-nanometer-thick walls. A similar investigation may be seen in the periodic cellular materials (PCMs) developed by Cellular Materials International. CMI's MicroTruss consists of a pyramidal core structure encapsulated between metallic face sheets and is lightweight yet sturdy enough to withstand ballistic impact.[6] In addition to shrinking metal structures, scientists are using a variety of means to dematerialize metals, including the creation of transparent iron.[7] **↗ Although such technologies are not yet applicable at a building scale, the emerging scientific emphasis on structural lattices is exerting a reciprocal influence on contemporary architecture.**

This approach is supported by next-generation technologies in digitally fabricated metal. **➔ Joris Laarman Lab and Acotech's MX3D Metal process couples an industrial robot with a welding machine, making it possible to print wireframe metal structures in midair (see page 96).** The company plans to build the world's first digitally fabricated bridge, to be composed entirely of robotically placed steel, over a central waterway in Amsterdam. Unlike first-generation additive manufacturing, which utilized plastic or cornstarch to create thick, stratified objects, MX3D Metal produces a thin skeletal framework using minimal material and without the need for supporting structures. Conceptually, this direction is reflected in other gossamer works, such as Lalvani Studio's expanded steel X-Tower and Jinil Park's Drawing series wireframe furniture (see page 84).

Complementing the newfound ephemerality of metal is its capacity for responsiveness. Shape memory alloys (SMAs) and smart foils enable materials to behave like

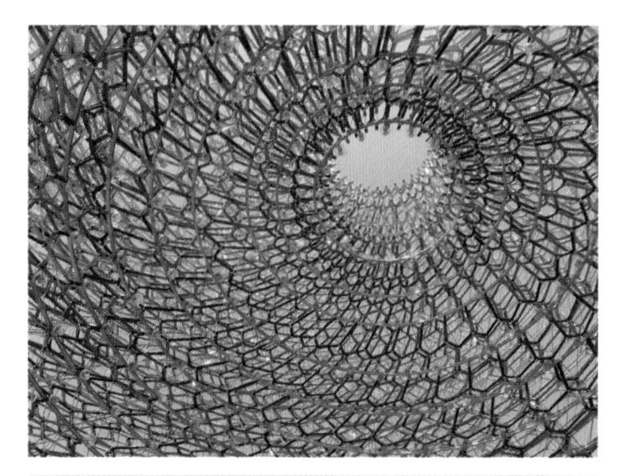

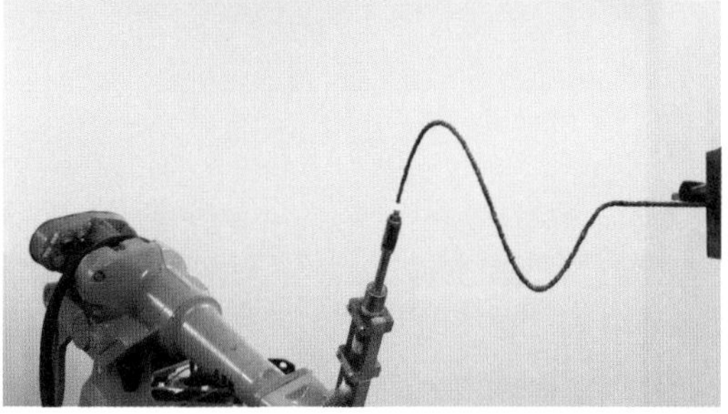

machines, changing their size and position in reaction to environmental factors. For example, Lift Architects' Air Flow(er) is a zero-energy ventilated surface that uses nickel-and-titanium-based wires to open up with increasing ambient temperatures (see page 80). **➔ Studio Roosegaarde's Lotus consists of smart aluminum mylar foils that curl back when a visitor approaches (see page 92).** The Exo process, developed by DOSU Studio Architecture, employs shape-shifting thermobimetal strips to build lightweight shell-based tower structures without the need for connecting hardware (see page 86).

Future metal technologies will be increasingly influenced by resource considerations. For example, the unabated global demand for power has inspired the development of reduced-energy manufacturing methods, such as Flash Bainite steel and flash-formed iron. The flash process enables the direct reduction of ore with hydrogen or natural gas, eliminating the need for traditional coke ovens (see page 88). Liquidmetal, a collection of alloys that may be thermoformed like plastic, are manipulable at much lower temperatures than standard industrial metals.[8] From a material-sourcing standpoint, resource criticality has become an important issue. According to a *YaleNews* article by Kevin Dennehy, "Many of the metals needed to feed the surging global demand for high-tech products, from smart phones to solar panels, cannot be replaced, leaving some markets vulnerable if resources become scarce."[9] These profound concerns will shape a new era for metal, which will do more with less and therefore manifest a lighter footprint. From a design perspective, metal will continue to reflect strength and beauty, but at a more human scale.

---

1 These are general timelines and differ based on geographic region. I discuss them here for symbolism purposes rather than in-depth historical analysis.

2 Annette LeCuyer, *Steel and Beyond: New Strategies for Metals in Architecture* (Basel: Birkhaüser, 1999), 8.

3 Michael F. Ashby, *Materials and the Environment: Eco-Informed Material Choice*, 2nd ed. (Oxford: Butterworth-Heinemann, 2012), 22.

4 Katy Barnato, "Steel demand 'evaporating at unprecedented speed,'" CNBC, October 28, 2015, http://www.cnbc.com/2015/10/28/steel-demand-evaporating-at-unprecedented-speed.html.

5 Ashby, *Materials and the Environment*, 16.

6 See accessed January 16, 2016, http://www.cellularmaterials.net.

7 Ralf Röhlsberger et al., "Electromagnetically induced transparency with resonant nuclei in a cavity," *Nature* 482 (February 9, 2012): 199–203, http://www.nature.com/nature/journal/v482/n7384/full/nature10741.html.

8 Camille Haley, "Liquidmetal: Redefining Metals for the 21st Century," NASA, October 26, 2005, http://www.nasa.gov/vision/earth/technologies/liquidmetal.html.

9 Kevin Dennehy, "For metals of the smartphone age, no Plan B," *YaleNews*, December 2, 2013, http://news.yale.edu/2013/12/02/metals-smartphone-age-no-plan-b.

THERMALLY REGULATED VENTILATION SYSTEM

Air Flow(er)

The Air Flow(er) is an active ventilating surface. The zero-energy apparatus consists of an assemblage of flowerlike modules that open as the ambient air temperature rises. The Air Flow(er) references the principle of thermonasty, a phenomenon that describes the ability of plants such as the yellow crocus flower to move in response to temperature changes.

Designed by Lift Architects, the system employs shape memory alloy (SMA) as its active component. The alloy can be deformed below its trigger temperature of approximately 80 °F (27 °C), but above this threshold, it rapidly returns to its original shape. In this way, SMA wires have the capacity to open the ventilating modules when heated. Once the air temperature drops below 60 °F (16 °C), the wires relax and allow the panels to close.

Air Flow(er) has three module types based on typical building aperture applications: a single-layer window, a double-layer envelope, and a roof vent. All of these systems operate without electricity.

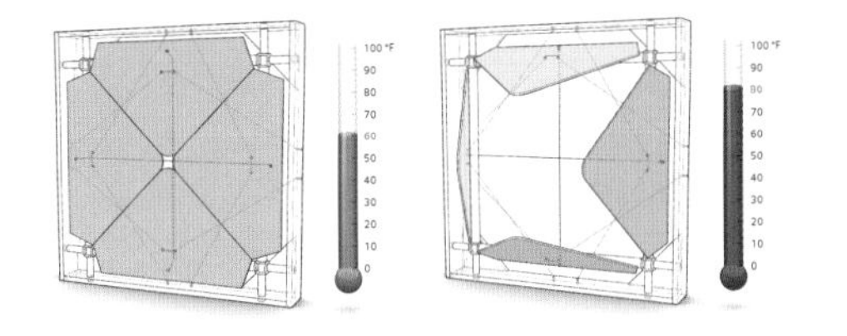

CONTENTS

Nickel- and titanium-based SMA wires, connector tubes, wire mesh screen, gusset plates, frame, elastic cords, barrel crimp, four-way connectors

APPLICATIONS

Vents, windows, ventilated building envelopes, thermally responsive divider walls

TYPES / SIZES

Roof vent, aperture, double-skin facade system

ENVIRONMENTAL

Energy-independent system, passive ventilation

LIMITATIONS

Because petals are opaque, closed modules reduce light and view penetration

FUTURE IMPACT

Expanded development and application of zero-energy, autoresponsive facade and mechanical systems

COMMERCIAL READINESS

●○○○○

CONTACT

Lift Architects
448 Broadway, #2, Cambridge, MA 02138
646-483-9731
www.liftarchitects.com
info@liftarchitects.com

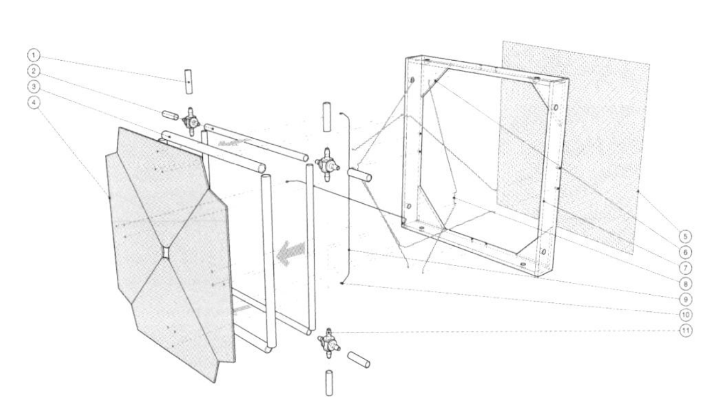

MUNICIPAL COMMUNICATIONS BEACON

Crisscross Signal Spire

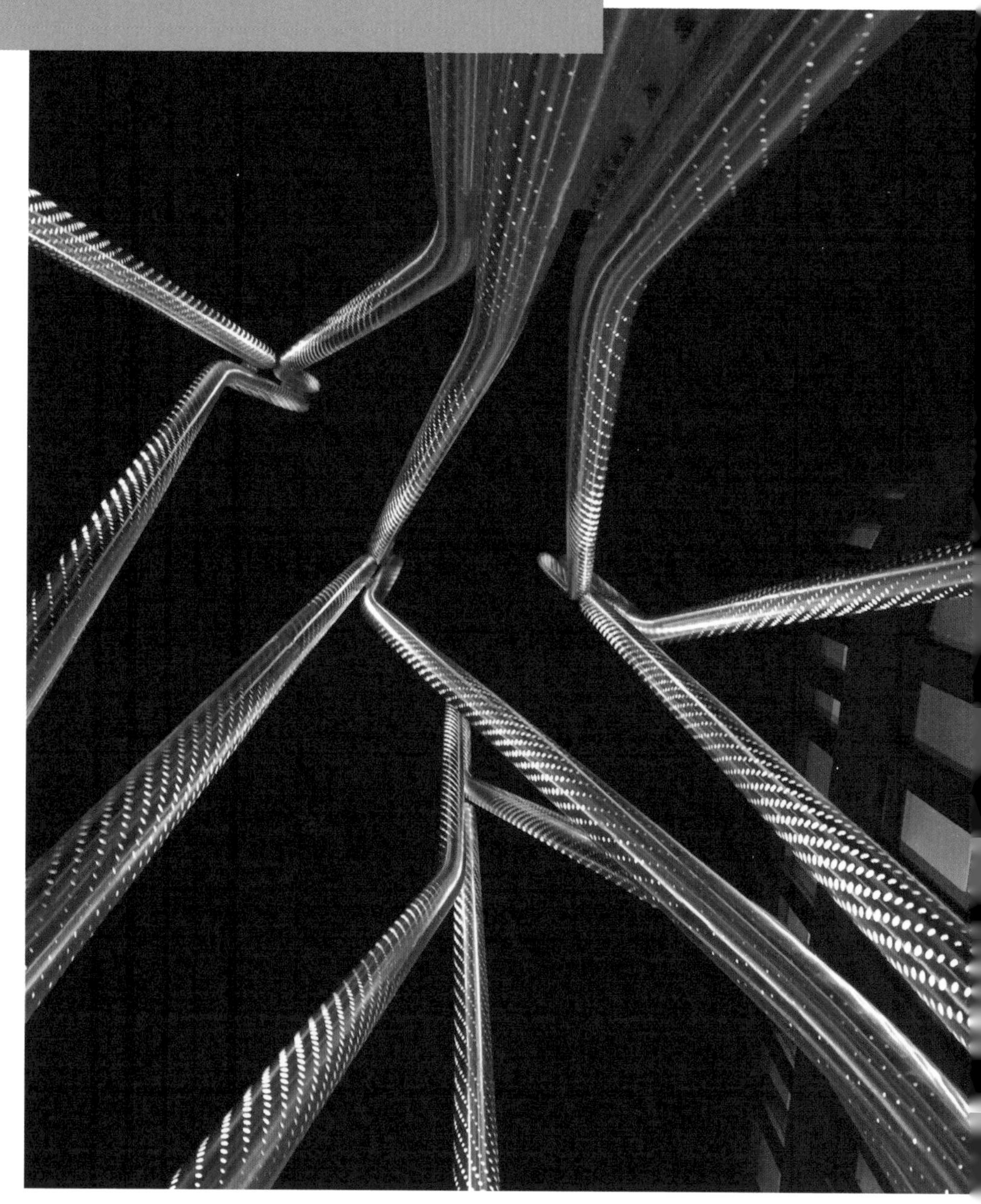

PRODUCT

The Crisscross Signal Spire is a twenty-first-century signal tower composed of internally illuminated vertical tubes. Situated in Boston's historic Dudley Square, the sculptural beacon embodies the function and symbolism of clock towers and church spires in a contemporary form. The spire acts as a digital interface for public communications and employs real-time lighting effects to broadcast valuable information from Boston's online Citizens Connect system. The signal tower conveys different types of data based on three distinct light patterns, which it emits through the graduated perforations in the tubes. One pattern marks time, akin to a digital hourglass; another indicates MBTA transit activity; and a third reveals 311 calls logged in the city's database. In this way, the spire becomes a sophisticated instrument for visualizing the various pulses of the city.

CONTENTS

Perforated steel tubes, LED lighting, digital communications equipment

APPLICATIONS

Communications displays, urban mapping devices, public sculpture

TYPES / SIZES

Vary

FUTURE IMPACT

Public communication of invisible or otherwise inaccessible information, visualization of infrastructural systems activity and intercity communication, contemporary interpretation of the traditional bell tower

COMMERCIAL READINESS

CONTACT

Höweler and Yoon Architecture LLP
150 Lincoln Street, 3A, Boston, MA 02111
617-517-4101
www.hyarchitecture.com
info@hyarchitecture.com

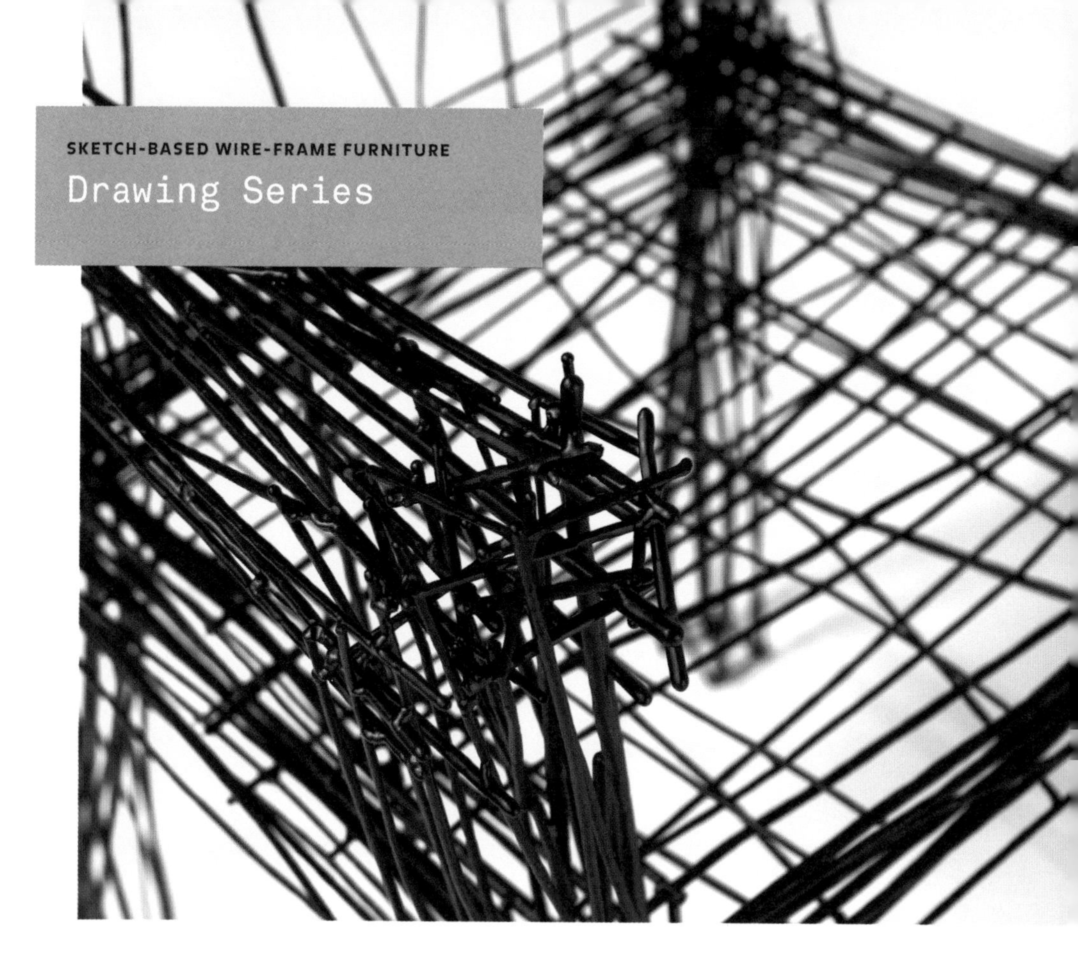

SKETCH-BASED WIRE-FRAME FURNITURE

Drawing Series

PRODUCT

When first encountering one of Korean designer Jinil Park's Drawing series, the viewer is likely to see the work as a two-dimensional sketch with loosely overlapping lines. However, closer inspection reveals corporeal furniture that occupies space and provides adequate structure for functional use. The concept was born when Park was sketching designs for new pieces, and he realized that the lines in the sketches could be translated directly into physical form.

The Drawing series consists of assemblies of carefully welded steel wires. According to the designer, the key point of the work concerns places where the lines become distorted, as these express the imperfect, emotional qualities of handcrafted drawings. The loose interplay of multiple imprecise lines not only reinforces the drawn aesthetic but also provides the necessary structural support unachievable with a single wire.

CONTENTS
Steel wires, paint

APPLICATIONS
Furniture

TYPES / SIZES
Six types of chairs, two tables, two lamps

LIMITATIONS
Not designed for mass production

FUTURE IMPACT
Loosening rigid preconceptions of drawing and making, expressing novel interpretations between drawing and making

COMMERCIAL READINESS

CONTACT
Jinil Park
B1, 31–5, Yangpyeong-dong 1-ga,
Yeongdeungpo-gu, Seoul 150-862,
South Korea
+82 1033909630
www.jinilpark.com
jiproject365@gmail.com

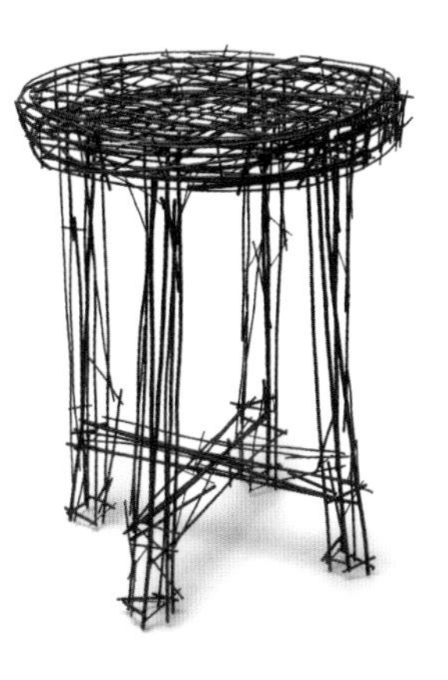

THERMOBIMETAL STRUCTURAL SHELL

Exo

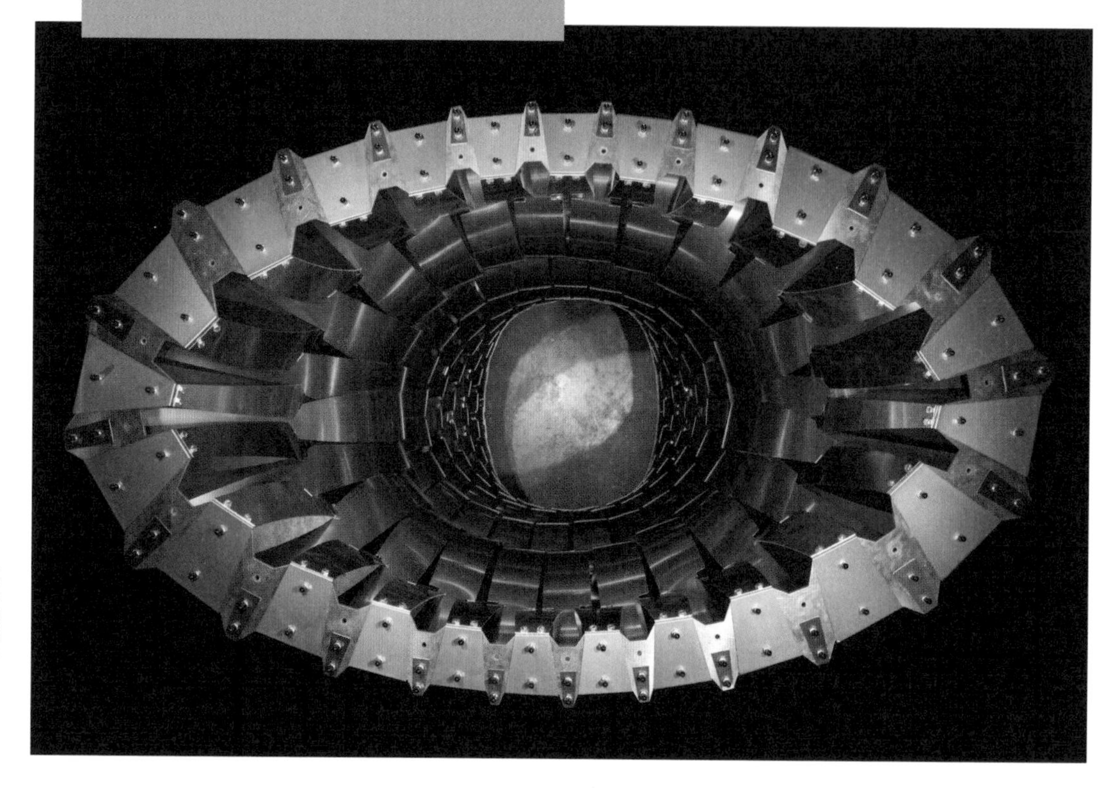

Exo is a vertical structure that makes use of the smart properties of thermobimetal, a coupling of two metal alloys that exhibit differing coefficients of thermal expansion. When heated, thermobimetal curls in a predictable direction, making it an appropriate material for transformable surfaces or constructions with minimal fasteners. To build Exo, designers at DOSU Studio Architecture heat thermobimetal strips to their activation temperature of 350 °F (177 °C) and hold them in place until they cool. After returning to their original flattened state, the alloy strips lock in place to create a pre-tensioned bow-beam. A collection of these bow-beams forms a robust and lightweight cylindrical shell that is reminiscent of a crustacean. This durable surface requires no connecting hardware, as Exo is held together entirely in tension.

CONTENTS

Laminated alloys of iron, nickel, and manganese; aluminum strips

APPLICATIONS

Streamlined construction of structural shells; pavilions, shelters, multifaceted surfaces

TYPES / SIZES

8'-tall (2.44-m) structure; 336 thermobimetal strips, 100 anodized aluminum strips

ENVIRONMENTAL

Fastener-free construction

TESTS / EXAMINATIONS

Mock-up exhibited at the Museum of Contemporary Art Santa Barbara, 2014

LIMITATIONS

Current Exo tower technology limited in size to a partial-scale prototype

FUTURE IMPACT

Resource-efficient, 4D building construction methods based on smart material properties

COMMERCIAL READINESS

●●○○○

CONTACT

DOSU Studio Architecture
8 Buggy Whip Drive, Rolling Hills, CA 90274
310-722-4458
www.dosu-arch.com
info@dosu-arch.com

GAS-REDUCTION IRON-MAKING PROCESS

Flash-Formed Iron

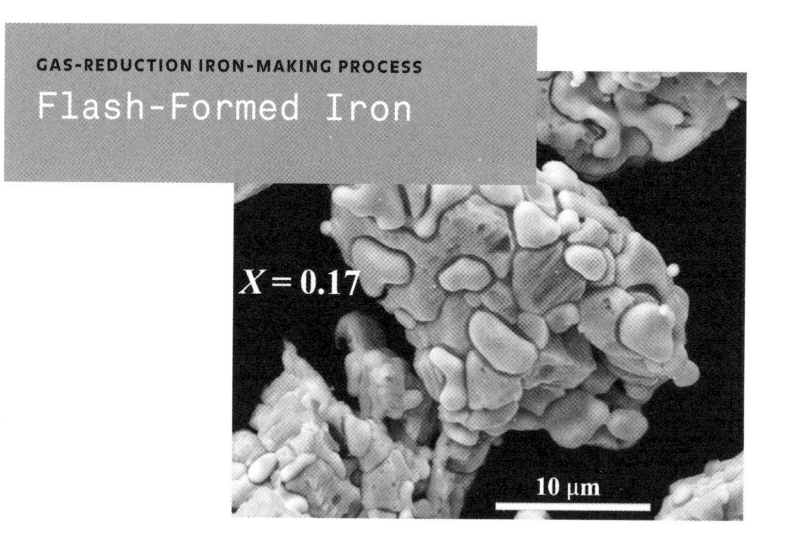

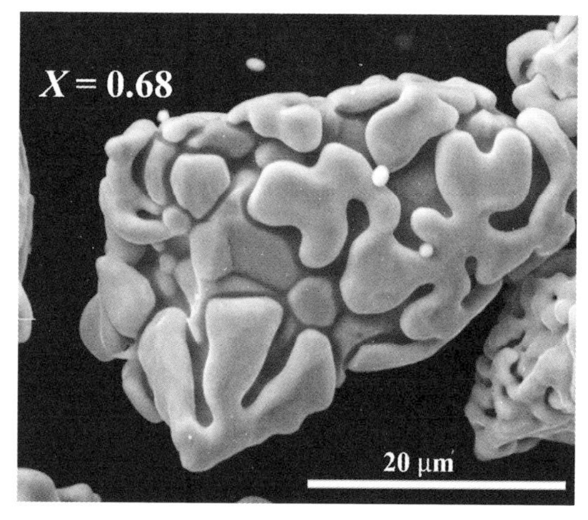

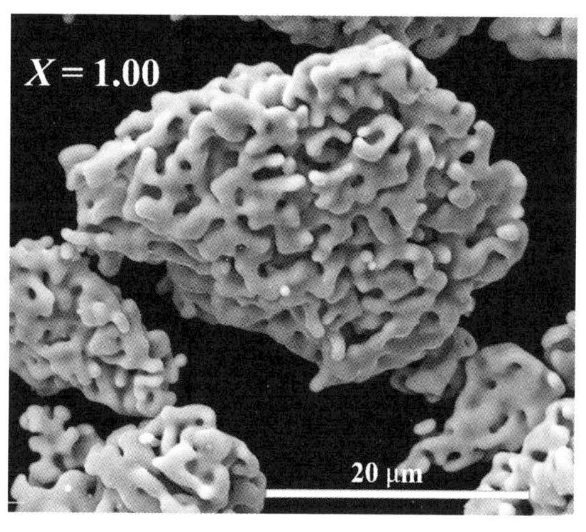

Although steel is the most recycled material globally, steel manufacture is still energy-intensive and exhibits a high carbon footprint. Scientists at the University of Utah have evaluated the conventional steel manufacturing process to improve its environmental performance. In the typical method of producing pig iron (crude iron) in steel making, a coke oven or blast furnace is used to process raw iron ore into sinter and pellets. High quantities of heat and forced air are required to melt the materials, resulting in a high embodied energy of 16–20 MJ/kg and a CO_2 footprint of 1.4–1.6 kg/kg for cast iron.

The researchers have developed a new flash iron-making process that avoids using coke, instead employing hydrogen or natural gas to reduce iron ore particles directly. This innovative method allows steel manufacturers to forgo the blast furnace entirely, shorten the processing time to seconds, and use smaller equipment for the same production volume—all of which lowers costs. Also, the new process is expected to reduce energy consumption by 20 percent and CO_2 emissions by 39 percent compared with traditional iron-making methods (the use of pure hydrogen would drop CO_2 emissions near zero).

CONTENTS
Iron ore, hydrogen or natural gas

APPLICATIONS
Iron and steel production

TYPES / SIZES
Vary

ENVIRONMENTAL
Significant reduction in embodied energy and CO_2 emissions

TESTS / EXAMINATIONS
Bench reactor commissioning, metallization determination, pilot plant design

LIMITATIONS
Hydrogen gas requires energy-intensive production

FUTURE IMPACT
Potential to improve the environmental performance of the world's second-largest industry

COMMERCIAL READINESS

CONTACT
University of Utah
Metallurgical Engineering &
Chemical Engineering
135 South 1460 East, Salt Lake City, UT 84112
801-581-6386
www.metallurgy.utah.edu
metal-info@lists.utah.edu

CONDUCTIVE MELTED ALLOY SURFACE

Form of Light Force Transmission

One method of creating a new material application begins with questioning the primary use of a commercial product. In Form of Light Force Transmission, Japanese designer Kouichi Okamoto reinterprets the conventional use of solder, a low-melting alloy used to connect fusible metals. Rather than join electrical wires together, Okamoto creates expansive surfaces with solder by methodically dripping the molten material over a large wood panel. Once the drips cool and fuse together, the designer attaches the solid alloy sheet to a vertical frame. The softly blurred mirror surface—reminiscent of pointillist imagery—is also a highly conductive plane, and Okamoto takes advantage of this by running an electrical current through the solder sheet to power connected lighting.

CONTENTS
Melted solder

APPLICATIONS
Wall or ceiling finish, power delivery

TYPES / SIZES
Installation including frame structure is 10' 6" × 6' × 4' 0" (3.2 × 1.8 × 1.2 m)

LIMITATIONS
Electrically conductive surface must be handled with caution

FUTURE IMPACT
Represents a creative approach to material misuse for generating novel applications

COMMERCIAL READINESS

CONTACT
Kyouei Design
1326-15 Kusanagi, Shimizu-ku,
Shizuoka 424-0886, Japan
+81 543470653
www.kyouei-ltd.co.jp
info@kyouei-ltd.co.jp

SMART FOIL RESPONSIVE SURFACE

Lotus

Lotus is a responsive surface made of pieces of smart foil that fold open in response to human proximity. When one approaches Lotus, hundreds of aluminum foils unfold themselves automatically, creating voids where the pieces fold away from each other. The smart foil, developed by Studio Roosegaarde and collaborating manufacturers, is composed of several thin layers of Mylar and aluminum that open when illuminated by the heat of interior lights. The responsive surface effectively tunes the materiality of the wall, transforming an opaque boundary into a visually permeable mesh.

Studio Roosegaarde has installed Lotus in two formats. Lotus Dome is a spherical vault made of hundreds of the "smart flowers," and Lotus 7.0 is a curved wall with vertically oriented smart foils.

CONTENTS

Ultralight aluminum and Mylar smart foils, lamps, sensors, software, and other media

APPLICATIONS

Responsive walls and ceilings, tunable apertures, interactive art

TYPES / SIZES

Lotus Dome is 10' × 10' × 6' 6" (3 × 3 × 2 m); Lotus 7.0 is a curved wall that is 13' × 6' 6" × 1' 7" (4 × 2 × 0.5 m)

ENVIRONMENTAL

Thermally reactive aperture system

FUTURE IMPACT

Material-based, self-calibrating systems without the need for additional energy inputs or mechanical components

COMMERCIAL READINESS

CONTACT

Studio Roosegaarde
Vierhavensstraat 52–54,
Rotterdam 3029 BG, Netherlands
+31 103070909
www.studioroosegaarde.net
mail@studioroosegaarde.net

ULTRALIGHT METALLIC CELLULAR MATERIAL

Microlattice

Inspired by the lattice structures of ambitious works of engineering, such as the Eiffel Tower and the Golden Gate Bridge, researchers at HRL Laboratories have developed one of the world's lightest materials. The new nickel-phosphorous substance consists of a cellular structure that resembles a multilayered space frame at microscale. Its incredibly low density—0.9 mg/cm³—makes it lighter than air and about one hundred times lighter than Styrofoam. It has an open volume of 99.99 percent, with 0.01 percent solid material deposited in the form of hollow tubes with a 100-nm wall thickness—or one thousand times thinner than a human hair.

Despite the material's low density, it exhibits high strength and elasticity, and can receive 50 percent strain without permanent deformation. According to the researchers, the microlattice architecture can be designed to be stiffer and stronger than foams and aerogels of the same density. It also has significant energy-absorption properties, making it suitable for vibration, acoustic, or shock energy-based damping.

CONTENTS

93 percent nickel, 7 percent phosphorous

APPLICATIONS

Ultralight textiles, battery electrodes, sound and vibration damping

TYPES / SIZES

Hollow-tube lattice with 100-nm-thick coating, material density of 0.9 mg/cm³

ENVIRONMENTAL

Extremely high strength-to-weight ratio, efficient use of material resources

TESTS / EXAMINATIONS

Transmission electron microscopy, compression testing

FUTURE IMPACT

Proliferation of ultralight, resource-efficient materials inspired by macroscaled approaches to structural engineering

COMMERCIAL READINESS

●●○○○

CONTACT

HRL Laboratories, LLC
3011 Malibu Canyon Road, Malibu, CA 90265
310-317-5000
www.hrl.com
media@hrl.com

3D-PRINTED METAL STRUCTURES

MX3D Metal

PROCESS

MX3D Metal is a multiple-axis 3D printing process that creates metal structures in midair. Developed by Netherlands-based Joris Laarman Lab in collaboration with Acotech, the method employs an industrial robot cleverly combined with a welding machine, as well as software custom developed to drive this hardware. The MX3D robot can print using various metals, including steel, stainless steel, aluminum, bronze, or copper, without the need for supporting elements. Starting from an anchored point on a horizontal or vertical surface, the robot delivers molten metal in small increments, resulting in lines of solid metal that emerge in midair. MX3D Metal may be used to create a wide variety of custom-designed structures, including a steel bridge that Joris Laarman Lab has proposed to print over a central waterway in Amsterdam.

CONTENTS
Steel, stainless steel, aluminum, bronze, or copper

APPLICATIONS
Metal wire-frame structures and surfaces, scaffolds, sculptural objects

TYPES / SIZES
Vary

ENVIRONMENTAL
Reduces material waste, eliminates need for supporting structures or formwork

LIMITATIONS
Requires time for molten metal to cool in small increments

FUTURE IMPACT
Automation of metal-frame construction; cost-effective capability to produce geometrically complex, resource-optimized metal structures

COMMERCIAL READINESS

CONTACT
MX3D
Ottho Heldringstraat 3,
Amsterdam 1066 AZ, Netherlands
+31 (0) 615595687
www.mx3d.com
info@mx3d.com

WIND-RESPONSIVE ILLUMINATED MESH

Windscreen

PRODUCT

Windscreen is a lightweight architectural lattice composed of an array of small wind turbines. The wind-activated structure generates and consumes renewable energy, translating the speed of local breezes into fluctuating levels of illumination.

Installed temporarily on the south face of MIT's Building 54 in Cambridge, Massachusetts, Windscreen flickered and vibrated as it indexed the wind, making information that is otherwise invisible, visible. The installation emphasized the connection between form and performance while demonstrating a fresh approach to integrating renewable power generation into architectural envelopes.

CONTENTS

Vertical-axis wind turbines, LED lights, steel cable net, frame

APPLICATIONS

Renewable energy harvesting, illumination, public art

TYPES / SIZES

Five hundred wind-activated lanterns

ENVIRONMENTAL

Net-zero lighting powered by local wind

FUTURE IMPACT

Multipurpose application of renewable energy technology serving mechanical engineering and public art functions

COMMERCIAL READINESS

CONTACT

Höweler and Yoon Architecture LLP
150 Lincoln Street, 3A, Boston, MA 02111
617-517-4101
www.hyarchitecture.com
info@hyarchitecture.com

4

Wood and Biomaterials

> Designing the Forest for the Trees

> Wood is emblematic of humanity's conflicted attitude toward natural resources. Although timber is a renewable resource, deforestation — which eradicates forests, often in the service of other land uses — continues at a rapid pace. In the past five millennia, approximately 4.45 billion acres (1.8 billion hectares) of land have been deforested, with an average net loss of 890,000 acres (360,000 hectares) annually.[1] According to a NASA study, the current pace of forest clearing will lead to the eradication of the world's rainforests in a century.[2]

This resource crisis suggests that forests must be preserved as much as possible, and suitable replacements must be found for wood-based materials. Yet alternative materials like concrete, metal, glass, and plastic are less environmentally desirable, due to their comparatively high embodied energy, high carbon emissions, and nonrenewable nature. Thus, wood continues to be a preferred material from an environmental perspective, with intensifying demand for a variety of applications, including tall buildings.[3]

This demand for a reliable source of wood products has led to the growth of managed forests. Since 1990 the total area of planted forests has increased by 270 million acres (110 million hectares).[4] These forests are much more like agricultural crops than wilderness and are effectively engineered in response to market demand.[5] "Today, improved or superior trees are increasingly being planted and are in fact the norm in the United States," claims forest economist Roger A. Sedjo. "In many other countries, emphasis is on improving exotic species that have become the dominant commercially planted trees."[6] The field of industrial forestry thus views forests not as uncultivated regions, but as territories subject to design.

Other renewable resources are being cultivated to meet the growing demand for wood. These include bamboo, hemp, nutshells, and straw—as well as recycled lumber, cardboard, and newspaper. ↙ **Novel composites, material hybrids, and repurposed products that exhibit greater sensitivity to natural resources are also being developed at a growing rate.** A few of these also demonstrate innovative capabilities, like self-shaping, based on designers' exploitation of wood's inherent physical properties. As a result, new wood and biomaterial products reveal a changing landscape for renewable materials, both figuratively and literally, with significant consequences.

Designers and manufacturers embracing wood's distinctive material characteristics challenge decades of industrial efforts to produce a stable, homogeneous material. For example, wood is hygroscopic, which means it gains and releases moisture from the air, depending on the air's relative humidity. Many engineered wood products are intentionally designed to suppress hygroscopy by the incorporation of glues, oils, and water-resistant coatings.[7] However, architect Achim Menges has long been interested in exploiting wood's hygroscopy for important functions. For example, the HygroSkin pavilion makes use of climate-responsive wood "leaves" to regulate light and air between exterior and interior environments (see page 108). The thin plywood veneer petals respond to shifts in the ambient humidity, in a range between 30 and 90 percent, passively opening and closing based on absorbed

and released moisture. → **Designers also take advantage of the flexibility that solid wood exhibits when partially cut in a cross-grain direction, as seen in Dukta's textile-like wood surfaces and LaRo Design's Spring Wood furniture (see page 124).** Clever material hybridization also enables shape-shifting abilities in designs such as the Self-Folding Paper developed by British Columbia–based engineer Ata Sina (see page 120).

Nanotechnological research in wood has resulted in new materials with capacities that would have previously been unthinkable in traditional wood products. For example, ↘ **scientists at the U.S. Forest Products Laboratory have developed high-performance nanocellulose composites from wood.** Toughened by cellulose nanofibrils and cellulose nanocrystals (CNC), the new material exhibits strength equivalent to Kevlar and is transparent.[8] Lead researcher Alan Rudie is currently investigating applications in composite windshields and other robust optical surfaces. Georgia Tech's Institute for Electronics and Nanotechnology has employed cellulose nanocrystals in the creation of a new organic semiconductor-based solar cell (see page 110). Unlike silicon-based photovoltaic technologies, this cell more closely approximates the material origins of biological photosynthesis.

The growing market for biomaterials is encouraging the development of alternative cellulosic materials. An example is the cardoon, or artichoke thistle, a plant common in Greece and neighboring countries. By combining cardoon fibers and bioresin, designer Spyros Kizis has made a new biocomposite material for furniture and other applications (see page 106). The Neptune grass that frequently covers Mediterranean beaches is another natural resource with a previously untapped use. Based on research conducted at the Fraunhofer Institute for Chemical Technology ICT, the seaweed is now processed for use as blown thermal insulation in building cavities (see page 112). Interest in the broad range of cellulose-based substances has motivated the exploration of materials and processes not typically seen in the design trades. Xylinum Cones, for example, are geometric modules made by shaping living bacterial cellulose into rigid objects (see page 132). **↓ MIT researchers have also explored the use of chitosan-based hydrogels in water-based manufacturing. These hydrogels are suitable for creating delicately structured surfaces resembling the composition of biological tissues (see page 116).**

Proliferating explorations in wood and other biomaterials suggest a potential change in industrial priorities, enabling modern societies to return to a more ecologically responsible resource balance. As Institute for Local Self-Reliance cofounder David Morris explains in his article "The Once and Future Carbohydrate Economy," in the magazine *American Prospect*, up until two centuries ago industrialized societies relied primarily upon plant-based feedstocks, but have since become addicted to nonrenewable industrial resources.[9] Due to the untenable results, modern industry is beginning to reverse course. The newfound enthusiasm for a carbohydrate economy (one based on plant resources) has the potential to spur a third industrial revolution. "We may be changing the very material foundation of industrial economies," Morris says. "It is an exciting historical opportunity, but one we should approach with deliberation and foresight."[10] Increasing pressures on agricultural output threaten to escalate global conflicts related to renewable resources, and sensible policies must be established to guide industrial priorities. Despite the challenges ahead, the carbohydrate economy could catalyze advantageous, nontraditional connections between designers, manufacturers, forest engineers, and farmers. The opportunity to cultivate a global industrial garden is upon us, and we should venture forth with care and sensitivity toward maintaining an optimal ecological balance.

1 *State of the World's Forests* (Rome: Food and Agriculture Organization of the United Nations, 2012), 9.
2 See accessed January 16, 2016, http://earthobservatory.nasa.gov/Features/Deforestation/tropical_deforestation_2001.pdf.
3 See an example of the reThink Wood campaign here, which describes wood's environmental advantages yet does not address deforestation concerns: accessed January 16, 2016, http://www.awc.org/docs/default-source/default-document-library/building-tall-with-wood-backgrounder.pdf.
4 "World deforestation slows down as more forests are better managed," Food and Agriculture Organization of the United Nations, September 7, 2015, http://www.fao.org/news/story/en/item/326911/icode/.
5 Roger A. Sedjo, "Regulation of Biotechnology for Forestry Products," in *Regulating Agricultural Biotechnology: Economics and Policy*, eds. Richard E. Just, Julian M. Alston, and David Ziberman (New York: Springer, 2006), 664.
6 Ibid.
7 Johan Liljencrants, "Wood impregnation and hygroscopy," accessed January 17, 2016, http://www.fonema.se/hygro/woodhyg.htm.
8 See accessed January 17, 2016, http://www.fpl.fs.fed.us/research/highlights/highlight.php?high_id=283.
9 David Morris, "The Once and Future Carbohydrate Economy," *American Prospect*, March 20, 2006, http://prospect.org/article/once-and-future-carbohydrate-economy. According to Morris, "As late as 1870, wood provided 70 percent of the nation's energy."
10 Ibid.

ARTICHOKE THISTLE FRP CHAIR

Artichair

PRODUCT

The artichoke thistle is an edible plant that has been harvested in central and western Mediterranean regions since ancient times. One of the most prevalent plant species in Greece, the artichoke thistle is well adapted to dry climates and has been identified as a promising natural resource for the local biofuel industry.

Seeking a fresh use for byproducts of the plant, designer Spyros Kizis has developed a biodegradable ecomaterial composed of the plant fibers and bio-resin—essentially a plant fiber-reinforced plastic. Kizis has used the material to create compression-molded components for a variety of interior furnishings, including a lounge chair, dining chair, and coffee table.

Because the new composite utilizes plant components that are not employed to make biofuels, the fabrication process may be creatively coupled with biofuel production to take full advantage of the natural resource. Since the furniture is inherently biodegradable, artichoke thistle seeds are intentionally interspersed within the reinforced material during fabrication. Buried pieces can thus germinate new plants.

CONTENTS

60 percent artichoke thistle fibers, 40 percent bioresin

APPLICATIONS

Molded furniture and housewares, product casings

TYPES / SIZES

Custom

ENVIRONMENTAL

Repurposed agricultural waste, low embodied energy

LIMITATIONS

Minimum thickness of 3/4" (2 cm)

FUTURE IMPACT

Potential partnerships among farming, energy, and furniture industries; resourceful utilization of agricultural by-products

COMMERCIAL READINESS

CONTACT

Kizis Studio
79 Greenwood Road, London E8 1NT, UK
+44 (0) 7554727374
www.spyroskizis.com
spyroskizis@gmail.com

CLIMATE-RESPONSIVE WOOD APERTURES

HygroSkin

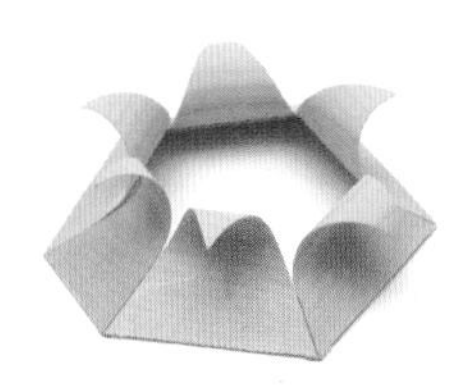

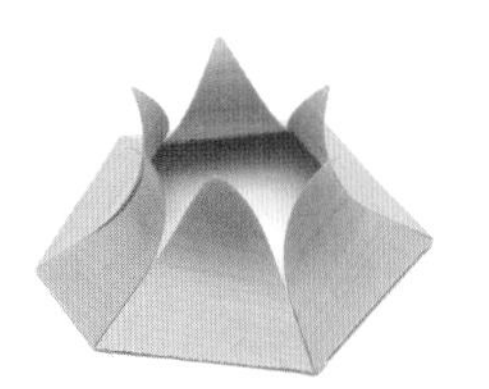

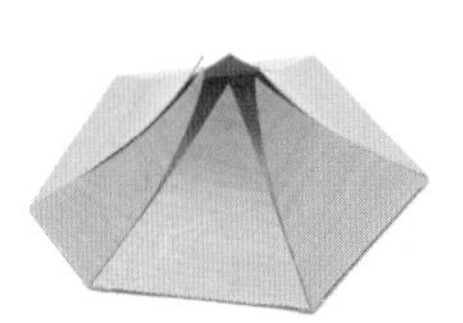

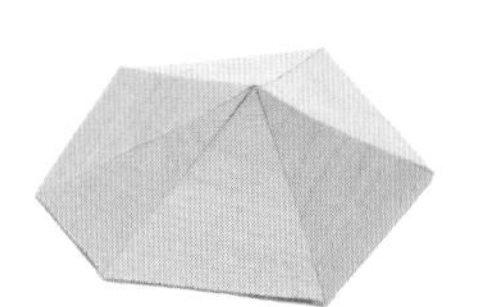

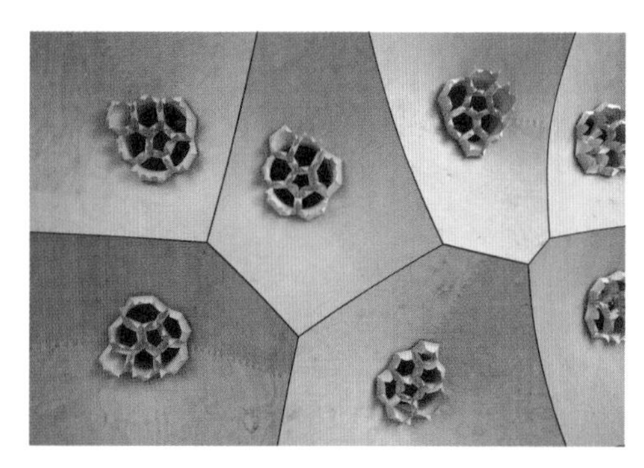

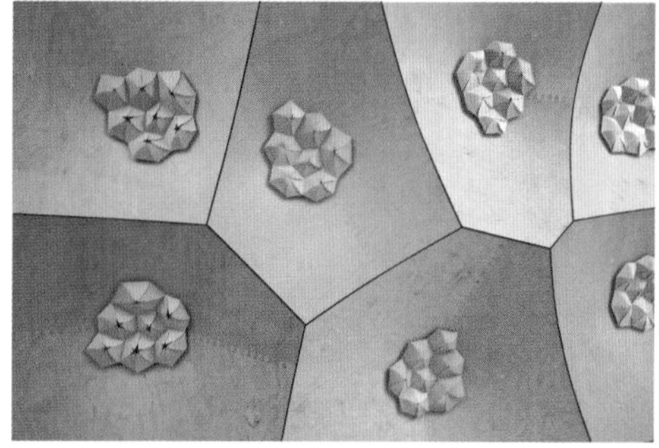

Although most environmentally responsive technologies rely on complex mechanical assemblies of inert materials, such as mechanical louvers or sunshades, Hygro-Skin demonstrates the reactive capabilities of a material itself. The hygroscopic behavior of wood, which causes it to swell and curl when moist, is the driving force behind this novel autonomic aperture system inspired by a biological precedent. Developed by Achim Menges, Oliver David Krieg, and Steffen Reichert for a demonstration pavilion, the HygroSkin approach is loosely based on the spruce cone's passive response to climate, in which its seed scales open and close based on changes in humidity.

Assemblies of delicate plywood-veneer scales are set within a robotically crafted envelope composed of concave plywood sheets. These hygroscopic scales react to a shifting humidity range between 30 and 90 percent, equivalent to the difference between clear and rainy weather in moderate climatic zones. In the pavilion, 1,100 scales are distributed among twenty-eight geometrically unique building components. As the petals adjust, they regulate the amount of direct light exposure, view access, and natural ventilation to the interior. Thus, this meteorologically sensitive behavior represents the bridging between environmental and spatial phenomena, all based on an entirely passive process. In HygroSkin, the material has become the machine.

CONTENTS
Plywood veneer, polymer substrate

APPLICATIONS
Responsive apertures, building envelopes, climate monitoring

TYPES / SIZES
28 unique geometries, 1,100 aperture scales

ENVIRONMENTAL
Opening that requires no energy or mechanical systems, renewable materials

TESTS / EXAMINATIONS
Demonstration pavilion installed as part of the permanent collection of the FRAC Centre, Orléans, France

LIMITATIONS
Operating humidity range of 30 to 90 percent

FUTURE IMPACT
Expanded research into material mechanics, development of responsive envelope systems that regulate climate with minimal or no energy or mechanical components

COMMERCIAL READINESS

CONTACT
Achim Menges Architect BDA
Klettenbergstrasse 24, Frankfurt 60322, Germany
+49 (0) 6995504310
www.achimmenges.net
mail@achimmenges.net

SOLAR CELL MADE FROM TREES

Nanocellulose Solar Cell

MATERIAL

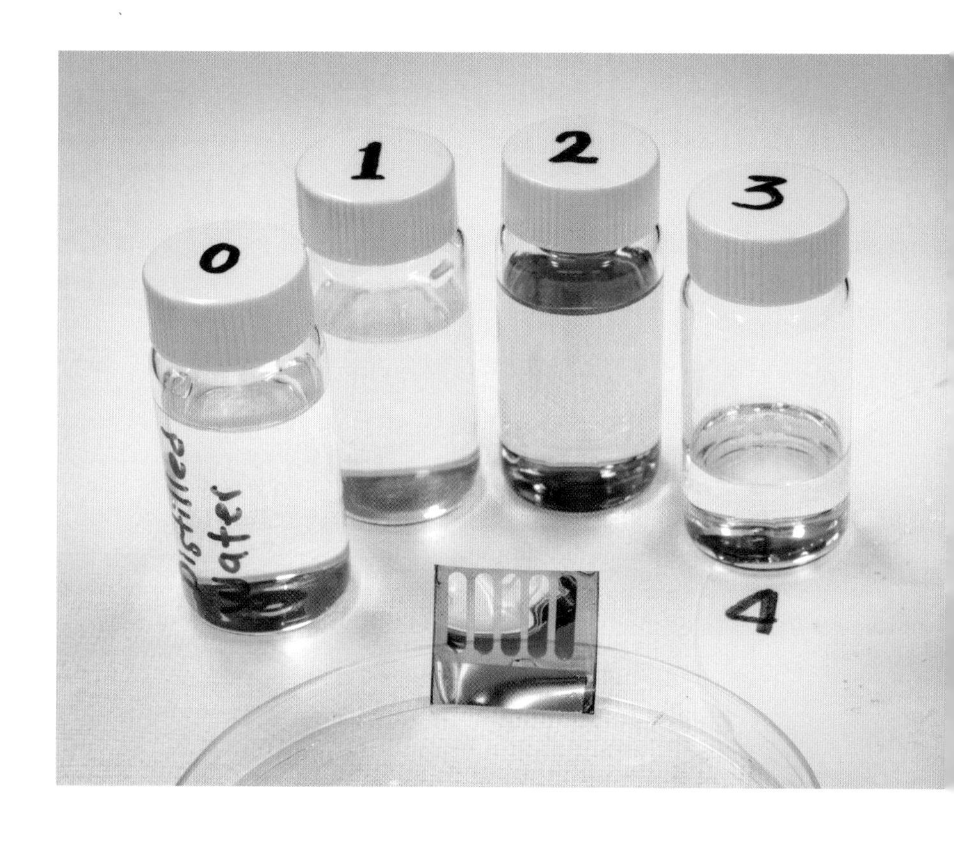

Although artificial solar cells are designed to emulate the natural process of photosynthesis, they could not be more materially different from the plant leaves they mimic. Typically made from silicon, plastic, and glass, conventional photovoltaic cells are energy-intensive to produce and difficult to disassemble and recycle.

Scientists at Georgia Tech's Institute for Electronics and Nanotechnology have developed a solar cell that more closely resembles the composition of tree leaves. The cell consists of two layers, both of which are derived from plant materials: an organic semiconductor and a transparent cellulose nanocrystal (CNC) substrate. The power conversion efficiency of the cell is just under 3 percent, which—although not yet competitive with the 10 percent levels of glass- or polymer-based cells—is unparalleled for cells made from renewable materials. Also, the device is easy to recycle at the end of its functional life: after the cell is immersed in water, the CNC substrate will dissolve, enabling easy separation of the cell's components.

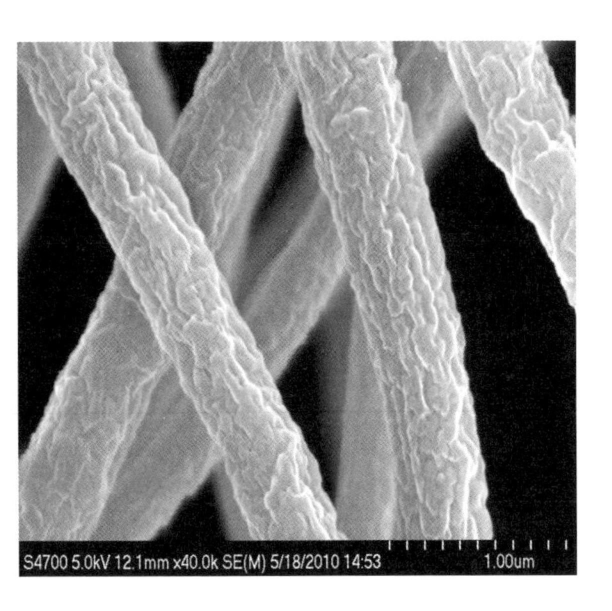

CONTENTS

CNC substrate, organic semiconductor

APPLICATIONS

Electricity generation from direct and diffuse light

TYPES / SIZES

Vary

ENVIRONMENTAL

Renewable-energy harvesting with biobased materials, designed for easy disassembly and recycling

TESTS / EXAMINATIONS

Current density–voltage, distilled-water immersion tests

LIMITATIONS

Power conversion efficiency limited to 2.7 percent

FUTURE IMPACT

Transformation of renewable-energy industry from a technical-nutrient basis to a renewable-resource basis

COMMERCIAL READINESS

●○○○○

CONTACT

**Georgia Institute of Technology
Institute for Electronics and Nanotechnology
345 Ferst Drive NW, Atlanta, GA 30318
404-894-5100
www.ien.gatech.edu
info@ien.gatech.edu**

SEAWEED INSULATION

NeptuTherm

Throughout much of the year, Mediterranean beaches are littered with clumps of *Posidonia oceanica* seaweed, also known as Neptune grass. The plant often winds up in landfills, yet this widely abundant natural material has many positive characteristics that make it too valuable for the trash heap. Researchers at the Fraunhofer Institute for Chemical Technology ICT noted the seaweed's high insulating value, in addition to its mold resistance and low flammability, and speculated about its use as a building product.

They developed a method of processing the material whereby they break up the clumps and cut them into short fibers. This loose fill is suitable for blown insulation applications in building cavities such as walls, roofs, and ceilings. In this function, the *Posidonia* fibers exhibit excellent rot resistance as well as the ability to absorb and release water vapor without diminished insulating capacity. Furthermore, the Fraunhofer process—which is now employed by NeptuTherm to make a commercial product—does not require any chemical additives.

CONTENTS

***Posidonia oceanica* plant fibers**

APPLICATIONS

Insulation for building roofs, walls, ceilings, and foundations

TYPES / SIZES

Blown insulation composed of $^{9}/_{16}$–$^{3}/_{4}$"-long (1.5–2 cm) fibers

ENVIRONMENTAL

Widely available in some locations, nontoxic renewable resource, low-energy processing

TESTS / EXAMINATIONS

Thermal tests conducted by the Fraunhofer Institute for Building Physics IBP, Stuttgart, Germany

LIMITATIONS

Requires special blower designed for *Posidonia oceanica* fibers

FUTURE IMPACT

Replacement of toxic, petroleum-derived insulating materials with inexpensive, renewable, and fully biocompatible resources

COMMERCIAL READINESS

CONTACT

Fraunhofer Institute for Chemical Technology ICT
Joseph-von-Fraunhofer Strasse 7, Pfinztal 76327, Germany
+49 72146400
www.ict.fraunhofer.de

REVERSE-PROCESSED LUMBER

NewspaperWood

NewspaperWood reverses the production process of making paper by making a wood-like material out of newspapers. NewspaperWood feedstocks include both pre- and postconsumer recycled paper from printing presses in Eindhoven, Netherlands, where the material is made. By utilizing a plasticizer- and solvent-free adhesive, the manufacturer ensures a trouble-free recycling process at the end of the product's functional life.

The striking visual quality of NewspaperWood is the result of its resemblance to wood grain. The manufacturer purposely cuts the many layers of glued newspaper that comprise the raw material to reveal dramatic curvilinear shapes reminiscent of timber products. Despite this close resemblance, it is evident upon close inspection that NewspaperWood is not a virgin lumber material. Nevertheless, it can be processed just like wood—including milling, cutting, and sanding.

CONTENTS

Newspaper, solvent-free adhesive

APPLICATIONS

Furnishings, interior finishes, decorative objects, automotive interiors

TYPES / SIZES

Custom sizes up to a maximum area of 15 × 6.7" (38 × 17 cm)

ENVIRONMENTAL

100 percent preconsumer recycled content, nontoxic adhesive enables 100 percent recycling

LIMITATIONS

For interior use only, not intended as a building-scale structural alternative to wood

FUTURE IMPACT

Upcycling potential to create desirable products from waste, integration of material streams to create successful cradle-to-cradle approaches

COMMERCIAL READINESS

CONTACT

NewspaperWood BV
Hallenweg 1E, Eindhoven,
Noord-Brabant 5615 PP, Netherlands
+31 (0) 408200069
www.newspaperwood.com
info@newspaperwood.com

WATER-BASED FABRICATION OF BIOMATERIAL STRUCTURES

Pneumatic Biomaterials Deposition

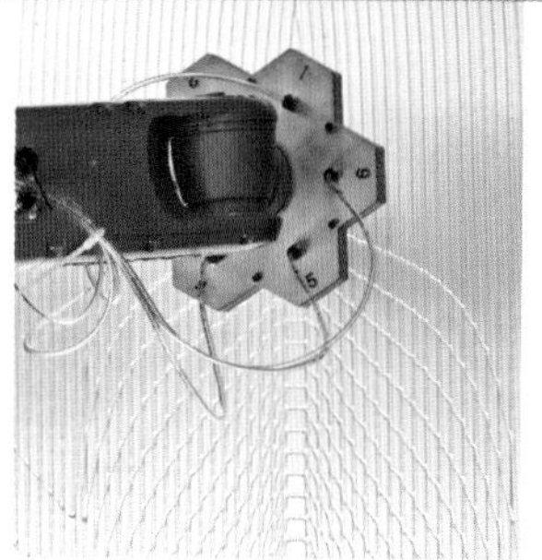

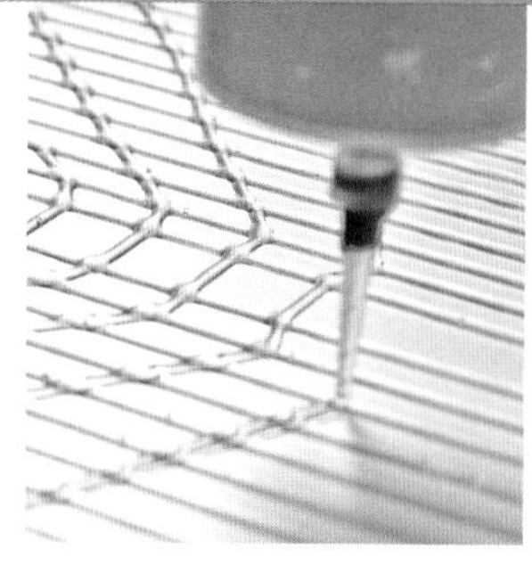

PROCESS

Researchers at MIT have realized a technique for printing gel-based structures modeled after the principles of biological construction. The water-based additive manufacturing method is driven by a robotic arm, equipped with a pneumatic deposition apparatus composed of multiple nozzles and syringes, that enables precise control over a variety of material gradients. The system prints hydrogels and other gel-based composites with a viscosity range of 500 to 50,000 cPs (the equivalent of motor oil to ketchup), and various resins, pastes, clays, and polyvinyl alcohols may also be incorporated into the gel feedstock. The materials cure after printing, changing from gels to solids. This material transformation allows designers to produce flexible, macro-scale architectural surfaces or wearable prostheses based on the heterogeneous cellular characteristics of biological structures.

CONTENTS
Chitosan-based hydrogels and cellulose composites

APPLICATIONS
Tissue scaffolds, prosthetic wearable devices, temporary architectural structures

TYPES / SIZES
Vary

ENVIRONMENTAL
Enables use of biocompatible, recyclable substances

LIMITATIONS
Viscosity range from 500 to 50,000 cPs

FUTURE IMPACT
Designed surfaces and structures that emulate the composition of biological tissues

COMMERCIAL READINESS
● ○ ○ ○ ○

CONTACT
Mediated Matter Group
Massachusetts Institute of Technology
Media Lab
75 Amherst Street, Room E14-433B,
Cambridge, MA 02139-4307
617-324-3626
matter.media.mit.edu
ked03@media.mit.edu

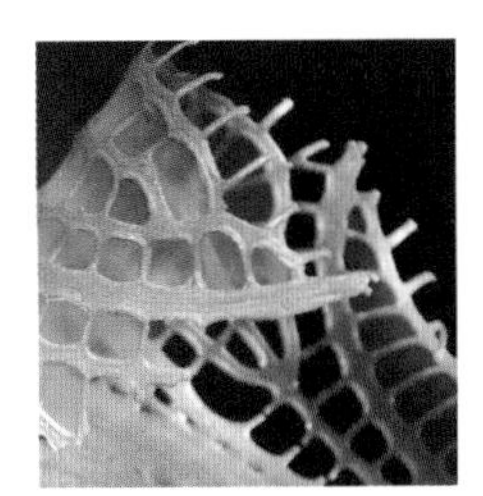

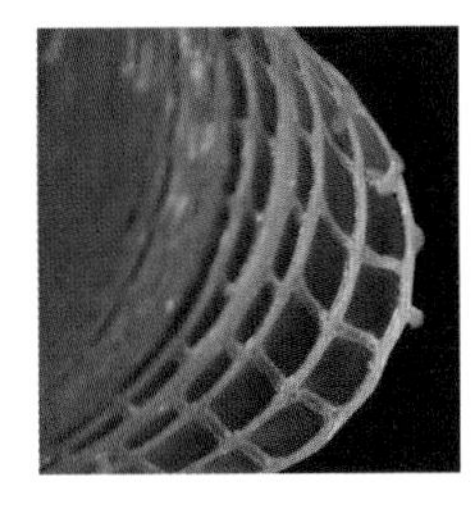

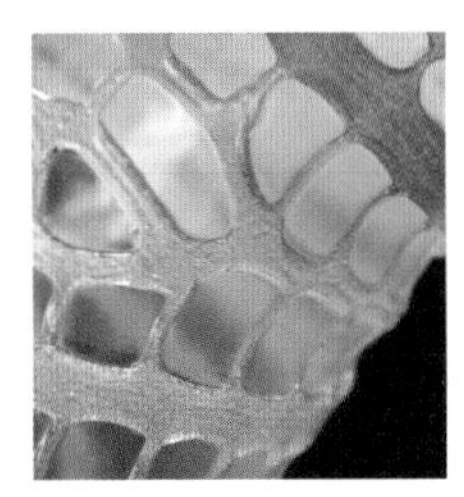

BAMBOO AND ROPE STRUCTURAL SYSTEM

Rising Canes

Rising Canes is a structural system made exclusively of bamboo and natural fiber ropes. Developed by Beijing design firm Penda for Beijing Design Week 2015, Rising Canes is an entirely modular, ecologically responsible, and expandable system for building construction.

Penda sought to explore the structural potential of bamboo because it is two to three times stronger by weight than steel and is rapidly renewable, and it has a small carbon footprint and can purify air near the shoot. Since ropes are used for structural connections, no nails, screws, or other mechanical fasteners are required in the Rising Canes system. Moreover, the use of rope leaves the bamboo struts undamaged and thus easily repositionable when new structural configurations are desired.

CONTENTS

Bamboo, rope

APPLICATIONS

Building structures, outdoor trellises

TYPES / SIZES

Vary

ENVIRONMENTAL

100 percent renewable and biocompatible materials, designed for disassembly

TESTS / EXAMINATIONS

Installation constructed for Beijing Design Week 2015

LIMITATIONS

Currently limited to a single story in height, with testing underway for taller structures

FUTURE IMPACT

True cradle-to-cradle building framing systems

COMMERCIAL READINESS

CONTACT

Penda
No. 13 Nafu Hutong, Dongcheng District, Beijing 100009, China
www.home-of-penda.com
info@home-of-penda.com

TRANSFORMABLE POLYMER COMPOSITE PAPER

Self-Folding Paper

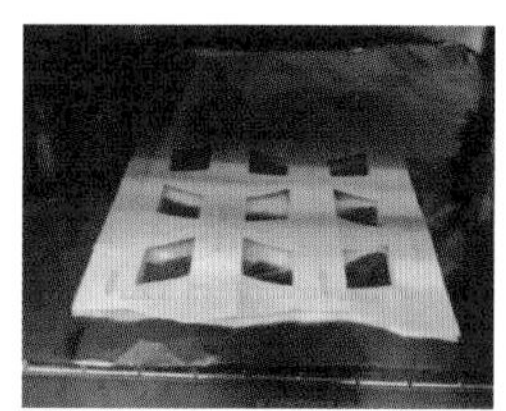

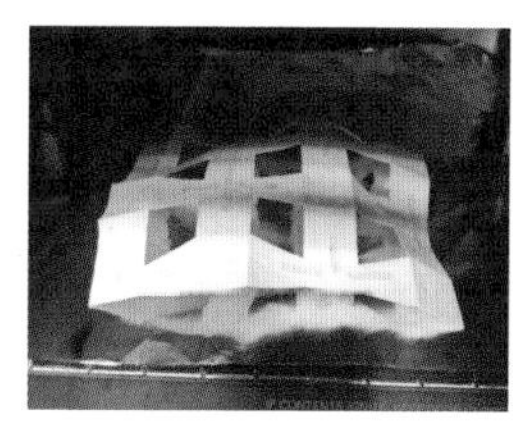

Self-Folding Paper is a paper surface that transforms when heated, automatically generating intricate multidimensional structures. Labeled thermocatalytic metafolds by mechanical engineer Ata Sina, who developed the technology at the University of British Columbia, the shape-shifting material is created by first making small computer-driven cuts and creases in a sheet of paper. Thermoplastic polymers are attached to the precut and precreased paper, which is then inserted into an oven at a temperature of 230 °F (110 °C) for ten to twenty seconds. As the polymers heat up, they shrink and lift the paper into various angles, turning it into a predetermined three-dimensional shape.

Sina declares that Self-Folding Paper will transform the packaging industry, because folded paper is light, strong, inexpensive, and more environmentally friendly than typical plastic packing materials. Other potential applications include noise and heat insulation, folding beds, step stools, toys, and do-it-yourself pop-up books.

CONTENTS

90 percent paper, 10 percent polymer

APPLICATIONS

Decorative objects, furnishings, acoustic surfaces, heat and noise insulation, packaging

TYPES / SIZES

Up to 12 × 12" (30 × 30 cm)

LIMITATIONS

Currently fabricated in a limited size of one square foot (0.093 m²)

FUTURE IMPACT

Lightweight transportability for flat-pack applications with on-site self-fabrication

COMMERCIAL READINESS

CONTACT

Joaters
2238 West 22nd Avenue, Vancouver, British Columbia, Canada V6L 1L7
604-600-7638
www.joaters.com
atasina@gmail.com

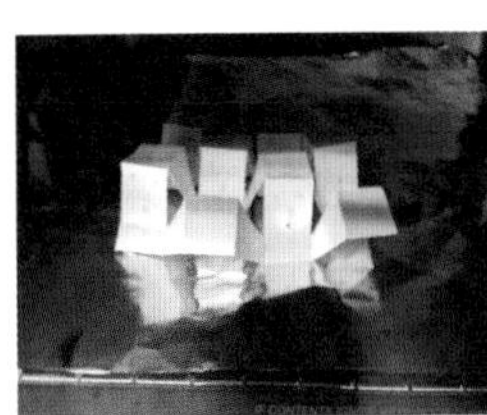
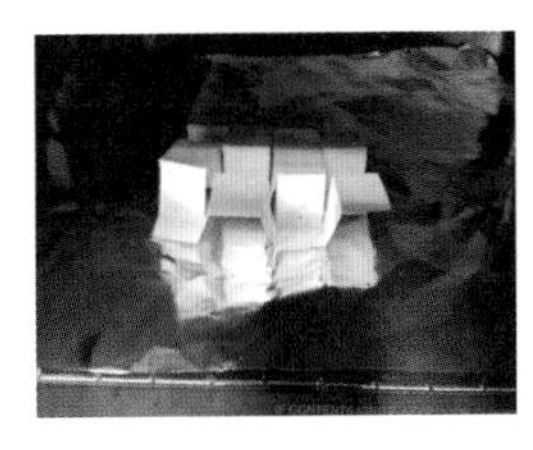
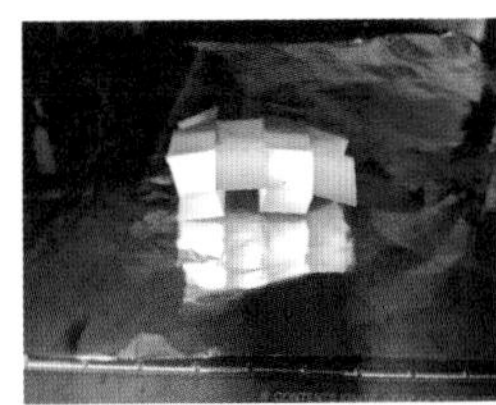

PRESSED SAWDUST FURNITURE

Shavings

Sawdust and other types of sawmill residues are typically treated as undesirable by-products of the timber industry and are often incinerated or dumped in landfills. Although there are many formats of engineered lumber that make use of sawdust and other wood fines, such as shavings and waste chips, the presence of such by-products is typically downplayed or disguised in finished furniture and millwork.

Design firm Producks Design Studio has embraced sawdust as a primary, conspicuous material in a line of furniture called Shavings. The designers mix sawdust with resin and press it into a mold to form a solid block that functions as a tabletop surface or a stool seat. The wood legs for the pieces are cleverly designed to interlock with the sawdust mixture to create a durable structural bond. This solid intersection is ensured by inserting the supports into the mold before casting, forming an integral joint.

CONTENTS

70 percent sawdust, 30 percent resin, oak or laminated birch

APPLICATIONS

Furniture, interior surfaces

TYPES / SIZES

Table 39 3/8 × 19 11/16" (100 × 50 cm), stool 13" (33 cm) diameter × 15 3/4" (40 cm) high

ENVIRONMENTAL

Functional repurposing of waste material

FUTURE IMPACT

Enhanced resource efficiency and creativity, new conspicuous uses for industrial by-products from the lumber industry

COMMERCIAL READINESS

CONTACT

Producks Design Studio
58 Gordon Street, Ramat Hasharon 47263, Israel
+972 (0) 523632611
www.pro-ducks.com
info@pro-ducks.com

FLEXIBLE INCISED SOLID WOOD

Spring Wood

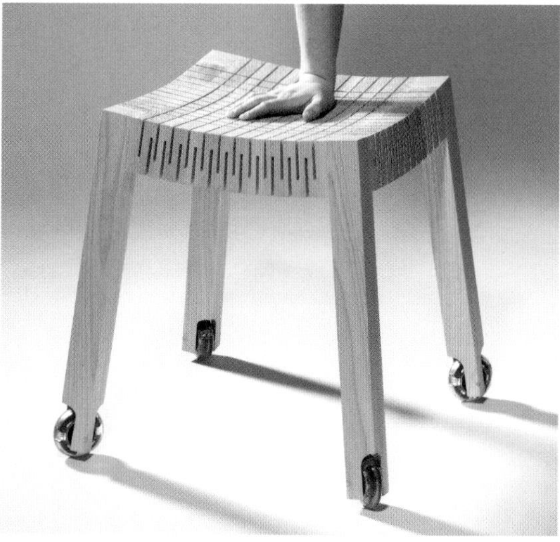

PRODUCT

Spring Wood makes use of strategically located cuts to impart solid wood with flexibility. Modern techniques to bend wood, used to create bent plywood furniture, include steaming and lamination. In LaRo Design's furniture collection, ash is precisely incised in a cross-grain direction to create two-sided, flexible wooden blocks. The collection includes four types of seating: the Original, the Paperclip, Restless Legs, and a three-seater bench called Bridge. In each piece, the flexibility of the wood provides a level of comfort that is unexpected in a square, solid wood seat.

CONTENTS

100 percent ash

APPLICATIONS

Furniture, flexible surfaces

TYPES / SIZES

Original, Paperclip, and Restless Legs 14 1/8 × 14 1/8 × 17 3/4" (36 × 36 × 45 cm); Bridge (three-seater) 59 × 19 11/16 × 17 3/4" (150 × 50 × 45 cm)

ENVIRONMENTAL

Rapidly renewable resource

LIMITATIONS

For interior use only

FUTURE IMPACT

Use of incision-based processing to convert rigid materials into flexible, ergonomic products

COMMERCIAL READINESS

CONTACT

LaRo Design
Noordendijk 153, Dordrecht, South Holland 3311 RN, Netherlands
+31 (0) 643754398
www.carolienlaro.nl
info@carolienlaro.nl

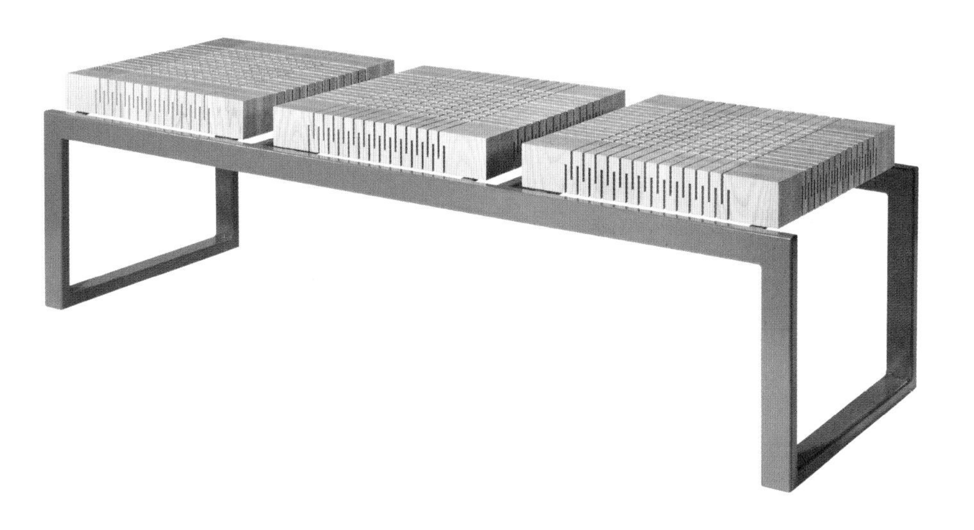

MOSSY WOOD ENCAPSULATED IN RESIN

Undergrowth

PRODUCT

Belluno, Italy–based design firm Alcarol aims to imbue its wood furniture with visual memories of the forests from which it originates. The firm sources lumber from fallen trees in the Italian Dolomites mountains, which feature a rich undergrowth of brightly colored mosses and lichens. The designers intentionally cut the wood in a way that preserves its natural, rough edges, which are covered with these chlorophyll-infused organisms. They then cast these plants in clear resin to maintain this colorful state and impart an aqueous visual effect.

Undergrowth furniture pieces are self-supporting structures with edges of embedded mossy bark. This unexpected internal contrast between the smooth planks and the irregular edges creates a dialogue between a found natural phenomenon and a refined, human-crafted object. The panels of the console units are joined at the corners to resemble a single bent piece, and the resin is made in one uniform cast.

CONTENTS
Solid wood, mosses, transparent resin

APPLICATIONS
Furnishings, decorative panels

TYPES / SIZES
Customizable

ENVIRONMENTAL
Utilizes salvaged lumber from fallen trees

TESTS / EXAMINATIONS
Anti-UV degradation

LIMITATIONS
For interior use only

FUTURE IMPACT
Products that reveal their natural origins in creative ways, providing visual cues to their life cycles

COMMERCIAL READINESS

CONTACT
Alcarol
Nongole 200, Belluno BL 32100, Italy
+39 3807252536
www.alcarol.com
info@alcarol.com

LIVING CANOPY SYSTEM

Vermilion Sands

PRODUCT

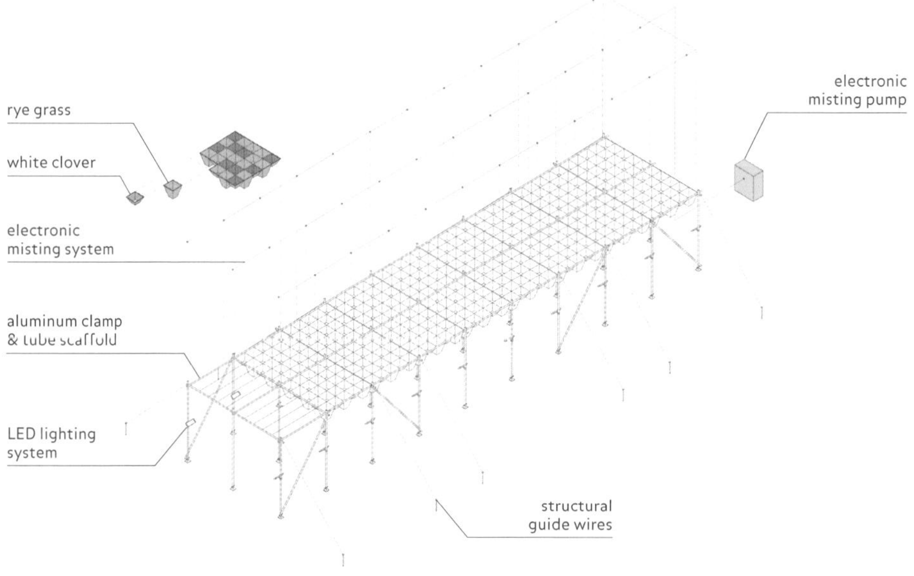

Vermilion Sands is a living ceiling composed of foliated geotextile modules. The system consists of a steel pipe structure and a grid of aircraft cables from which the modules are suspended. Each pyramid-shaped module is fabricated in a four-step process: First, light-gauge wire is bent to frame the desired shape. Second, geotextile fabric is sewn onto the frames. Third, a slurry mixture is sprayed onto the fabric in a hydroseeding process (a planting method with mulch and a seed slurry). Fourth, the hydroseeded module is grown in a nursery for one month before installation.

An array of misting nozzles is integrated into the canopy surface, which is designed for an outdoor public installation, to provide irrigation for the plants and reduce the ambient air temperature in hot climates. The resulting effect is that of an inverted landscape that combines abstract geometric forms with a living garden.

CONTENTS

75 percent hydroseed slurry (water, wood pulp, guar gum, and grass/clover seeds), 20 percent geotextile fabric, 5 percent 10-gauge steel wire

APPLICATIONS

Exterior shading, landscaping, thermal regulation

TYPES / SIZES

Perennial ryegrass module and white clover module; small module 24 × 24 × 12" (61 × 61 × 30.5 cm), large module 24 × 24 × 24" (61 × 61 × 61 cm)

ENVIRONMENTAL

Opportunity for living plant material to be incorporated into architectural surfaces, offers potential for adiabatic cooling

TESTS / EXAMINATIONS

Large-scale public installation at the 2015 Chicago Architecture Biennial

LIMITATIONS

Requires adequate light and moisture to support plant growth, requires continual maintenance

FUTURE IMPACT

Convergence of architectural surfaces and living plant systems

COMMERCIAL READINESS

CONTACT

Matthew Soules Architecture
529 Carrall Street, Vancouver, British Columbia, Canada V6B 2J8
604-568-1050
www.msaprojects.com
office@msaprojects.com

WOOD-BASED INSULATION

Wood Foam

PROCESS

Researchers at the Brunswick, Germany–based Fraunhofer Institute for Wood Research have developed a new wood-based insulation material that consists of foamed wood particles. Aiming to create a more environmentally effective substitute for conventional petroleum-derived insulating products, the researchers established a process to transform wood into aerated panels. They grind wood into a fine pulp and add gas to expand the particle soup into a bubbly foam. Next, the scientists allow the foam to harden into a lightweight base material, which they can process further into rigid panels or flexible mats. According to the team, wood foam exhibits performance and longevity comparable to traditional polymer foam insulation.

CONTENTS
Foamed wood particles

APPLICATIONS
Building insulation, packaging

TYPES / SIZES
Foams from a variety of tree species available

ENVIRONMENTAL
Renewable replacement for common petroleum-based insulating materials

TESTS / EXAMINATIONS
Laboratory tests

FUTURE IMPACT
Substituting renewable material surrogates for environmentally costly fossil fuel–derived products

COMMERCIAL READINESS

CONTACT
Fraunhofer Institute for Wood Research, Wilhelm-Klauditz-Institut WKI
Bienroder Weg 54E, Braunschweig 38108, Germany
+49 5312155208
www.wki.fraunhofer.de
volker.thole@wki.fraunhofer.de

GEOMETRIC CELLULOSE OBJECTS

Xylinum Cones

PROCESS

Xylinum Cones are the result of a production process that utilizes living organisms to grow geometric objects. Designers Jannis Hülsen and Stefan Schwabe work with bacterial cellulose, an organic material produced by particular kinds of bacteria, to explore the fabrication of new biotechnological materials and modular objects. Within a growth period of three weeks, each cellulose cone ripens within a suspended mold. At this time, different material properties can be added via simple chemical processes.

The shapes of the cones reference common natural geometries, such as reptile scales or flower seeds—which have analogs in building products such as roof tiles or clapboards. The designers' primary goals for the Xylinum Cones are to explore the architectural implications of living matter and to achieve a balance between geometric precision and natural diversity in biological objects.

CONTENTS
Bacterial cellulose

APPLICATIONS
Geometric objects

TYPES / SIZES
Vary

ENVIRONMENTAL
Biocompatible, biodegradable, recyclable

LIMITATIONS
Subject to natural degradation processes

FUTURE IMPACT
Development of novel methods for creating products composed of living materials

COMMERCIAL READINESS

CONTACT
Jannis Hülsen with Stefan Schwabe
Solmsstrasse 11, Berlin 10961, Germany
+49 1756859001
www.jannishuelsen.com
info@jannishuelsen.com

Plastic and Rubber

5

> A Second Plastic Age

> As the philosopher Roland Barthes
> declared in his oft-quoted
> essay on the material, "Plastic
> is the very idea of its infinite
> transformation."[1] One interpre-
> tation is that plastic, a malleable
> material capable of adopting seem-
> ingly limitless forms in iterative
> succession, is an intrinsically
> unbounded substance. Another,
> broader view is that plastic con-
> tinues to change over time and
> continues to redefine itself. As
> science writer Philip Ball states,
> "Since the advent of materials
> we call plastics, every age has
> reinvented these substances
> to reflect its own occupations."[2]

By the time Barthes's essay was published in 1957, plastic had already established a clear and far-reaching trajectory, with the burgeoning industrialization of many polymer technologies for military, business, and consumer uses. Since then, plastic has become ubiquitous, and it may be found in every part of our lives—including everyday objects, clothing, and the buildings we inhabit. We have now become so complacent about plastic's omnipresence that we fail to comprehend the profound set of changes currently underway: a widespread transformation that suggests the inception of a new era for plastic.

Plastic's first epoch was defined by its invention in the mid-nineteenth century, followed by a period of significant industrial development. By the 1930s and '40s, a polymer revolution was in progress, with the creation of many now-ubiquitous materials, such as neoprene, acrylic, PVC, polyurethane, polyethylene terephthalate (PET), Teflon, and fiber-reinforced plastic. The swift emergence of so many fascinating substances inspired many to theorize about their impact on the physical environment. In 1941 chemists V. E. Yarsley and E. G. Couzens announced the beginning of the Plastic Age. They believed that future generations would enjoy "a world of color and bright shining surfaces, where childish hands find nothing to break, no sharp edges or corners to cut or graze, no crevices to harbor dirt or germs."[3]

In a short time, plastic did indeed transform the world of consumer products in addition to many environments, which today exhibit the very qualities that Yarsley and Couzens envisioned. However, first-generation plastics have a poor environmental track record. These materials are derived from petroleum and are responsible for one hundred to five hundred million tons of CO_2 emissions annually (synthetic rubber, which has replaced many applications of natural rubber, is likewise synthetic and derived from oil). Many plastics contain red list chemical toxins and contribute widespread, pervasive contaminants, such as hormone-altering chemicals. Moreover, only 7 percent of plastics

are recycled, leaving the remainder to be burned, landfilled, or discarded thoughtlessly to pollute the world's oceans.[4]

Current polymer research and development points to several promising new directions for a second Plastic Age. According to bioengineer Don Ingber, "There is an urgent need in many industries for sustainable materials that can be mass produced." [5] Biopolymer production is now ramping up significantly, and bioplastics are increasingly employed as replacements for their oil-based counterparts. Biopolymers offer many advantages over fossil fuel–derived plastics: they utilize renewable and biocompatible feedstocks, are often recyclable and compostable, and can be programmed to biodegrade within certain time frames and under particular environmental conditions. Although the first biopolymers exhibited brittleness and were made from edible agricultural resources, like corn or soybeans, new versions are more mechanically robust and made from nonfood biomass like switchgrass and kenaf fiber. **← One of the most promising examples is derived from the chitin in discarded shrimp shells. Ingber and colleague Javier Fernandez's new chitosan bioplastic exhibits the strength and toughness of aluminum at only half the weight, and can be molded into virtually any form (see page 142).**

Next-generation plastics also exhibit smart properties, such as sensing, communication, response, and self-repair, imparting these materials with advanced capabilities that require no additional energy input. **→ The University of Michigan and NASA collaborated in creating a plastic that can self-heal after a high-impact puncture from a bullet or other projectile.**[6] The material contains a reactive fluid encapsulated between plastic sheets and is designed for aerospace applications in which atmosphere loss poses a critical threat. Other smart properties are evident in advanced materials such as liquid crystal elastomers, which are shape-responsive structures that gain stiffness when compressed, or photoluminescent bioplastic, a self-illuminating material made from the by-product of biodiesel fuel production (see page 154).

New methods also are underway to reuse existing petroleum-based plastics more resourcefully. The Ocean Cleanup project has the ambitious goal of collecting plastic waste from the world's oceans, starting with the North Pacific accumulation zone or "Great Pacific Garbage Patch." [7] Design office Studio Swine has launched a demonstration project in collaboration with fishermen called "Sea Chair." [8] They collect and sort marine plastic using fishing nets and melt it into furniture with a custom-built furnace. Creative efforts to salvage and refunctionalize waste polymers address many different waste streams. For example, at least five hundred billion disposable plastic bags are consumed annually, yet the average bag use is a mere twelve minutes.[9] **↗ Edinburgh-based Waël Seaiby upcycles discarded plastic bags into bowls, vessels, and various decorative objects called Plag (see page 156).**

Smart polymers, as well as bioplastics and upcycled plastics, demonstrate the fact that plastic is becoming more closely aligned with biological processes—in both physical and operational terms. The

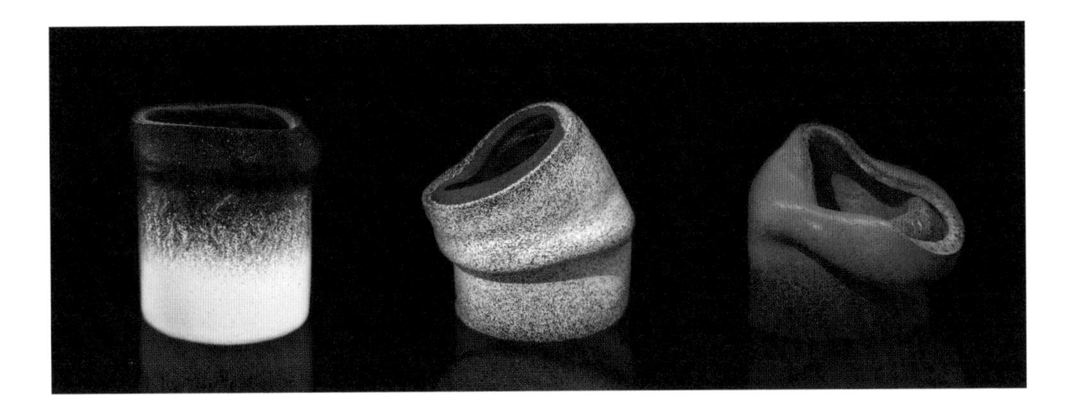

opportunity that plastic could be a bio-compatible, responsive, and even lifelike material would have been unthinkable in the first Plastic Age, an era that celebrated plastic's synthetic character. Plastic's growing similarity to life raises interesting questions in light of the material's typical use in simulating other substances. Such imitation will no longer be limited to color and texture, but also include dynamic performance. Bio-based plastics will also pose resource challenges to an increasingly crowded agricultural field that struggles just to feed the global population. Consequently, inedible crops, food waste, and agricultural by-products will serve as some of the most viable feedstocks for new plastics. Harnessing the technical-nutrient waste stream of petroleum plastics will also be critical to cleaning the environment and reducing demand for new material, inviting new connections between plastic recyclers and product manufacturers.

Despite plastic's shifting terrain, one aspect remains unchanged: it offers no proscriptive path to application, but rather invites open speculation about its optimal form and behavior. In this sense, it is the ultimate design material and beckons us to figure out how to apply it. When asked how we should design with plastics, architect Billie Faircloth replies: "I dare you to try."[10]

---

1 Roland Barthes, *Mythologies* (New York: Noonday Press,1972), 97.
2 Philip Ball, "The Plastic Proteus," in Billie Faircloth, *Plastics Now* (Abingdon, UK: Routledge, 2015), 331.
3 V. E. Yarsley and E. G. Couzens, "The Expanding Age of Plastics," *Science Digest* 10 (December 1941): 57–59.
4 According to Columbia University's Earth Institute, only 6.5 percent of American plastic is recycled, out of 33.6 million tons discarded annually. See Renee Cho, "What Happens to All That Plastic?" *State of the Planet* (blog), January 31, 2012, http://blogs.ei.columbia.edu/2012/01/31/what-happens-to-all-that-plastic/.
5 Don Ingber quoted in "Manufacturing a solution to planet-clogging plastics," Wyss Institute press release, Harvard University, March 3, 2014, http://wyss.harvard.edu/viewpressrelease/144/.
6 See accessed January 25, 2016, http://www.acs.org/content/acs/en/pressroom/presspacs/2015/acs-presspac-august-26-2015/self-healing-material-could-plug-life-threatening-holes-in-spacecraft-video.html.
7 Boyan Slat, "Why we need to clean the ocean's garbage patches," the Ocean Cleanup, December 15, 2016, http://www.theoceancleanup.com/blog/show/item/why-we-need-to-clean-the-oceans-garbage-patches.html.
8 See accessed January 25, 2016, http://www.studioswine.com/sea-chair/.
9 See accessed January 25, 2016, http://www.citizenscampaign.org/campaigns/plastic-bags.asp.
10 Faircloth, *Plastics Now*, 373.

3D-PRINTED POLYAMIDE SEAT

Biomimicry

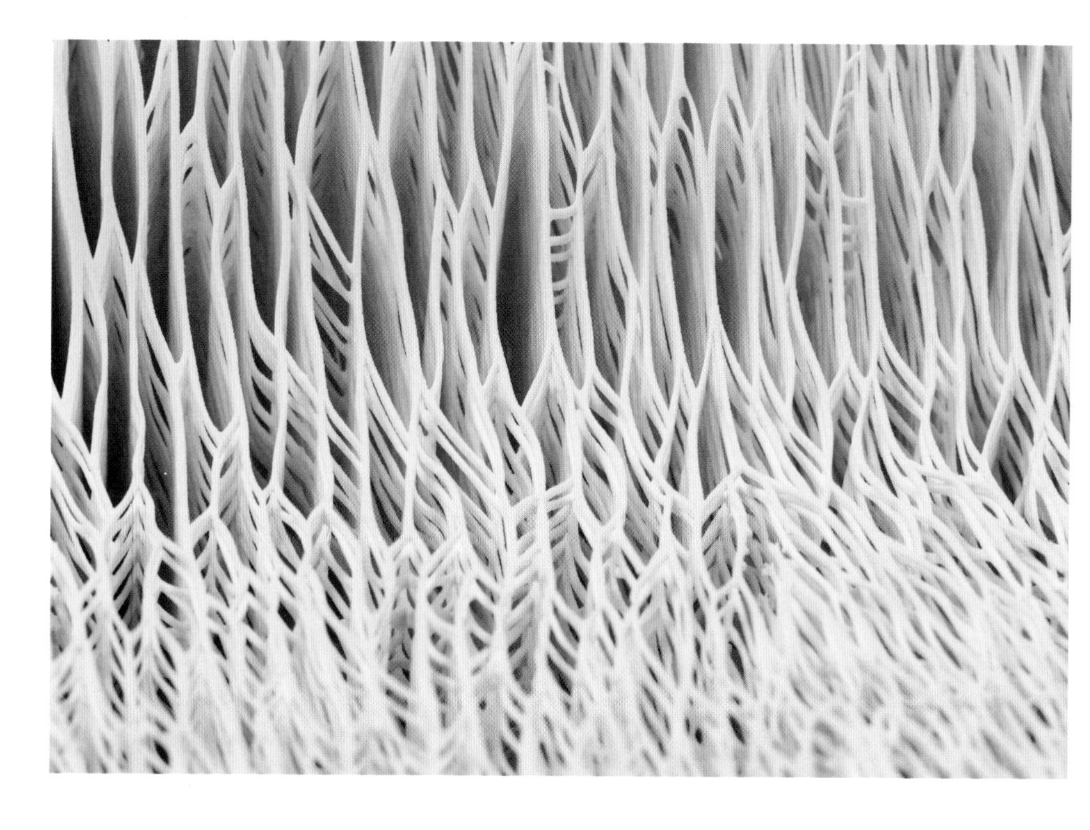

PRODUCT

With the aid of 3D printing, designer Lilian van Daal's Biomimicry chair represents a more sustainable alternative to the conventional furniture production practices of combining a variety of dissimilar materials with toxic adhesives. Based on the geometry of natural organisms, such as the strong yet lightweight structures of plant cells, van Daal's design embodies multiple material properties using a single substance, without the need for adhesives. By varying the structure in several places, the designer achieves the myriad characteristics of load-bearing construction, micro and macro support, ventilation, and skin—all using polyamide material. Unlike traditionally manufactured seating, the monomaterial soft seat is also readily recyclable at the end of its functional life.

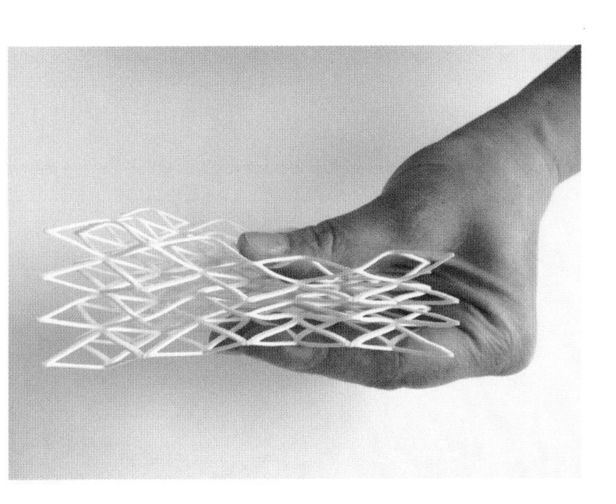

CONTENTS
Polyamide

APPLICATIONS
Furniture, decorative objects, lighting

TYPES / SIZES
Vary

ENVIRONMENTAL
Multifunctional monomaterial assembly eliminates need for multiple materials and adhesives, recyclable

LIMITATIONS
Must be treated for outdoor use

FUTURE IMPACT
Proliferating uses of monomaterial applications in products and environments, increasing material efficacy and recyclability

COMMERCIAL READINESS

CONTACT
Lilian van Daal
Marten van Rossemstraat 8,
Arnhem 6821 BA, Netherlands
+31 (0) 63382 5479
www.lilianvandaal.com
info@lilianvandaal.com

CHITIN-BASED BIODEGRADABLE POLYMER

Chitosan Bioplastic

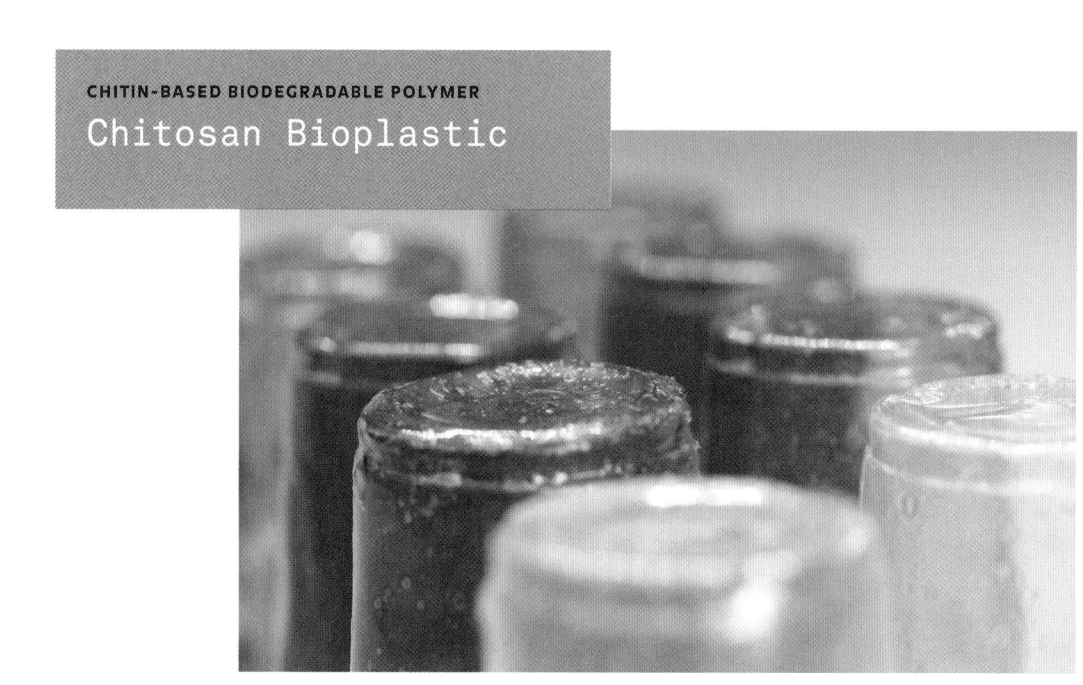

Scientists at Harvard University's Wyss Institute for Biologically Inspired Engineering have developed a new biopolymer that mimics the toughness, durability, and flexibility of natural insect cuticle. A biological composite with a structure akin to plywood, insect cuticle consists of layers of the polysaccharide polymer chitin and protein. The Wyss researchers extracted chitosan sugar from discarded shrimp shells and engineered a thin, transparent film with a strength equivalent to aluminum at only half the weight. The new Chitosan Bioplastic is biocompatible, biodegradable, easily molded into complex shapes, and inexpensive because of the widespread availability of shrimp waste. Because of these beneficial characteristics, the material is a potential replacement for conventional petroleum-based plastic in applications ranging from consumer products to medical uses.

CONTENTS
Chitosan, wood flour

APPLICATIONS
Thin films, packaging, bandages, sutures, mobile devices, toys

TYPES / SIZES
Wide variations in stiffness and coloration possible

ENVIRONMENTAL
Environmentally effective substitute for petroleum-derived polymers, biodegradable, biocompatible, recyclable

TESTS / EXAMINATIONS
Laboratory tests

LIMITATIONS
Chitosan's original molecular structure must remain intact for optimal performance

FUTURE IMPACT
Reduced reliance upon energy-intensive, fossil fuel–based plastics; reduced volume of nondegrading, environmentally polluting plastic waste

COMMERCIAL READINESS
●●○○○

CONTACT
Wyss Institute for Biologically Inspired Engineering
Harvard University
3 Blackfan Circle, Boston, MA 02115
617-432-7732
www.wyss.harvard.edu
info@wyss.harvard.edu

RECYCLED COFFEE-BASED POLYMER COMPOSITE

Çurface

MATERIAL

Although various methods to reuse coffee grounds exist, such as composting or converting coffee wastewater into energy, much of this waste goes to landfills.

Çurface is an innovative material developed by UK-based Re-worked and Smile Plastics that incorporates at least 40 percent recycled coffee grounds in combination with either bioresin or recycled plastic. The material has a richly toned, smooth surface and has the strength and durability of hardwood. Çurface is made to order either as a panel product or in bespoke moldings and is suitable for a range of applications, including appliance surrounds and tabletops.

CONTENTS

70 percent recycled coffee grounds, 20 percent resin, 10 percent jute

APPLICATIONS

Decorative panels, furniture, appliance casings, jewelry, decorative objects

TYPES / SIZES

Made to order, also available with recycled plastic instead of resin

ENVIRONMENTAL

80 percent recycled content, 20 percent bioresin; also available in 100 percent recycled content

TESTS / EXAMINATIONS

Laboratory tests

LIMITATIONS

Panels and furniture not recommended for exterior use

FUTURE IMPACT

Diversion of biomass waste from landfills and into new, biocompatible products

COMMERCIAL READINESS

CONTACT

Rosalie McMillan
London, UK
www.rosaliemcmillan.com
info@rosaliemcmillan.com

FRACTAL-BASED ENVIRONMENTAL SCREEN

Fibonacci's Mashrabiya

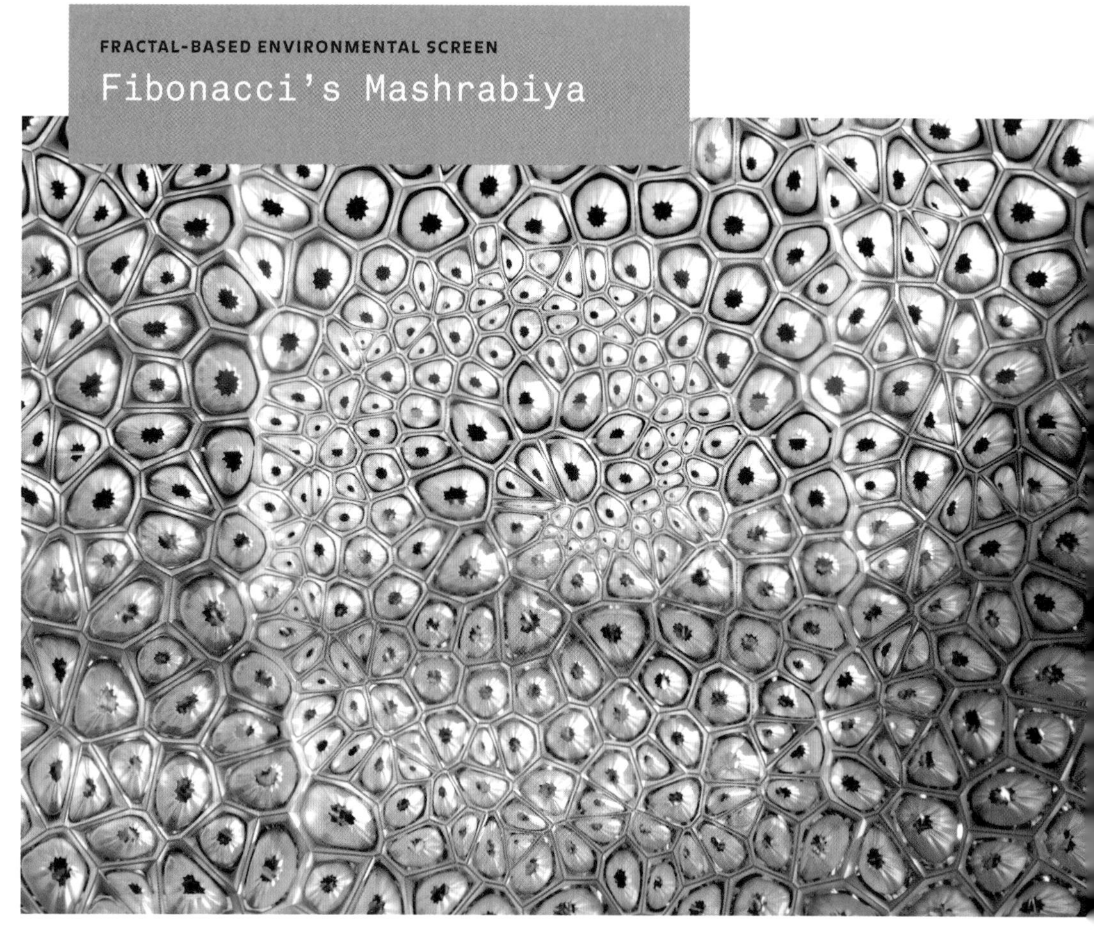

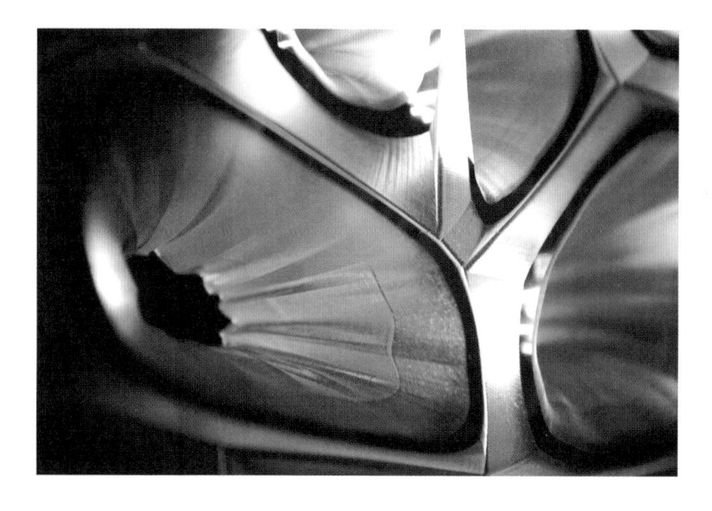

Mashrabiya is an Arabic term for a type of bay window with carved wooden latticework. A traditional element of Arabic architecture, the screen device permits the passage of air and access to views while maintaining the privacy of building occupants. Fibonacci's Mashrabiya is a contemporary interpretation of this historic building component and reenvisions the screen apparatus via computer algorithms and digital fabrication methods. Recognizing that the traditional mashrabiya enables a range of environmental effects through modulation of its pattern density and frame thickness, researchers at MIT employed fractal patterns derived from the Fibonacci sequence to manipulate air and light in new ways. The profile of the resulting CNC-milled acrylic panel varies with depth, imparting the screen with air-, light- and view-directing properties. As with the historic mashrabiya, the overall pattern and structure may be tuned to the microclimatic specificities of particular environments.

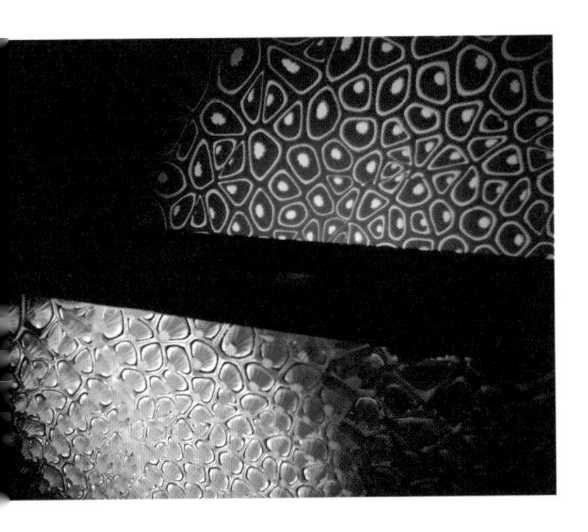

CONTENTS
CNC-milled acrylic

APPLICATIONS
Windows, screen walls, interior partitions

TYPES / SIZES
Vary

ENVIRONMENTAL
Patterns may be manipulated to control solar penetration

TESTS / EXAMINATIONS
Prototype installations at the Centre Pompidou, Paris, and Smithsonian Institution, Washington, DC

FUTURE IMPACT
Rediscovery and reinterpretation of historic architectural practices using contemporary technologies, creative engagement with aperture design and fabrication

COMMERCIAL READINESS

CONTACT
Mediated Matter Group
Massachusetts Institute of Technology Media Lab
75 Amherst Street, Room E14–433B, Cambridge, MA 02139-4307
617-324-3626
matter.media.mit.edu
ked03@media.mit.edu

HEXAGONAL THERMOFORMED PANELS

Hex

Hex thermoformed panels are fabricated from thermofoil molded to polyethylene. In conventional practice, thermofoil (a rigid plastic finish) is thermoformed to another material, such as medium-density fiberboard, for cabinetry. The thermofoil finish typically emulates natural wood grain and is therefore used to simulate the aesthetic of solid wood while reducing material cost.

Designer Andrea Valentini's Hex panels reimagine the thermoforming process with plastic laminates to create complex surfaces in a variety of colors and textures. Hexagonal molds enable the use of a single-ply surface that is lightweight and flexible, yet also sufficiently durable for interior surface treatments. The panels are also available in a wood grain pattern, which pokes fun at the traditional simulation of the material in thermofoil finishes.

CONTENTS

Laminated thermofoil, polyethylene

APPLICATIONS

Surface treatments for walls, ceilings, and cabinetry

TYPES / SIZES

Customizable colors, patterns, and sizes; material thickness 1/4" (0.635 cm)

LIMITATIONS

For use in interior and conditioned exterior environments

TESTS / EXAMINATIONS

Prototype installations at the Centre Pompidou, Paris, and Smithsonian Institution, Washington, DC

FUTURE IMPACT

Exploration of dimensionality in surface finishes to reduce required mass, substitution of faux finishes with novel surfaces

COMMERCIAL READINESS

●●●●●

CONTACT

Andrea Valentini
Wanskuck Mill, 725 Branch Avenue, #133, Providence, RI 02904
401-225-0289
www.andreavalentini.com
info@andreavalentini.com

HYBRID PLASTIC AND WOOD FURNITURE

Landscapes Within

PRODUCT

Landscapes Within is a collection of furniture that contains visible inclusions of natural materials, combining the traditional craft of basketry with modern materials and processes. Created by furniture designer Wiktoria Szawiel, the simple stools, chairs, and tables are intended to capture the spirit of the landscape in her native Poland.

Szawiel collected various grasses and other plants from the region and encapsulated them in translucent resin, which composes the structure in some pieces and acts as a pearly coating in others. The opalescent depth of the furniture imparts a mysterious visual quality, inviting the viewer to take part in a memorable process of discovery.

CONTENTS

Polyester, epoxy, and polyurethane with rattan, wicker, or wood inclusions

APPLICATIONS

Furniture, vessels

TYPES / SIZES

Objects up to 60" (150 cm) long and 2" (5 cm) thick

ENVIRONMENT

Ability to incorporate repurposed materials

LIMITATIONS

For interior use only

FUTURE IMPACT

Potential for upcycled product streams, multidimensional products that reveal material inclusions

COMMERCIAL READINESS

CONTACT

Wiktoria Szawiel
Rua General Norton de Matos, 92 Murtal, Parede, Lisbon 2775-132, Portugal
+35 1920218181
www.wiktoriaszawiel.com
wiktoria.szawiel@gmail.com

FLEXIBLE POLYMER NETWORK WITH LIQUID CRYSTAL ORDERING

Liquid Crystal Elastomer

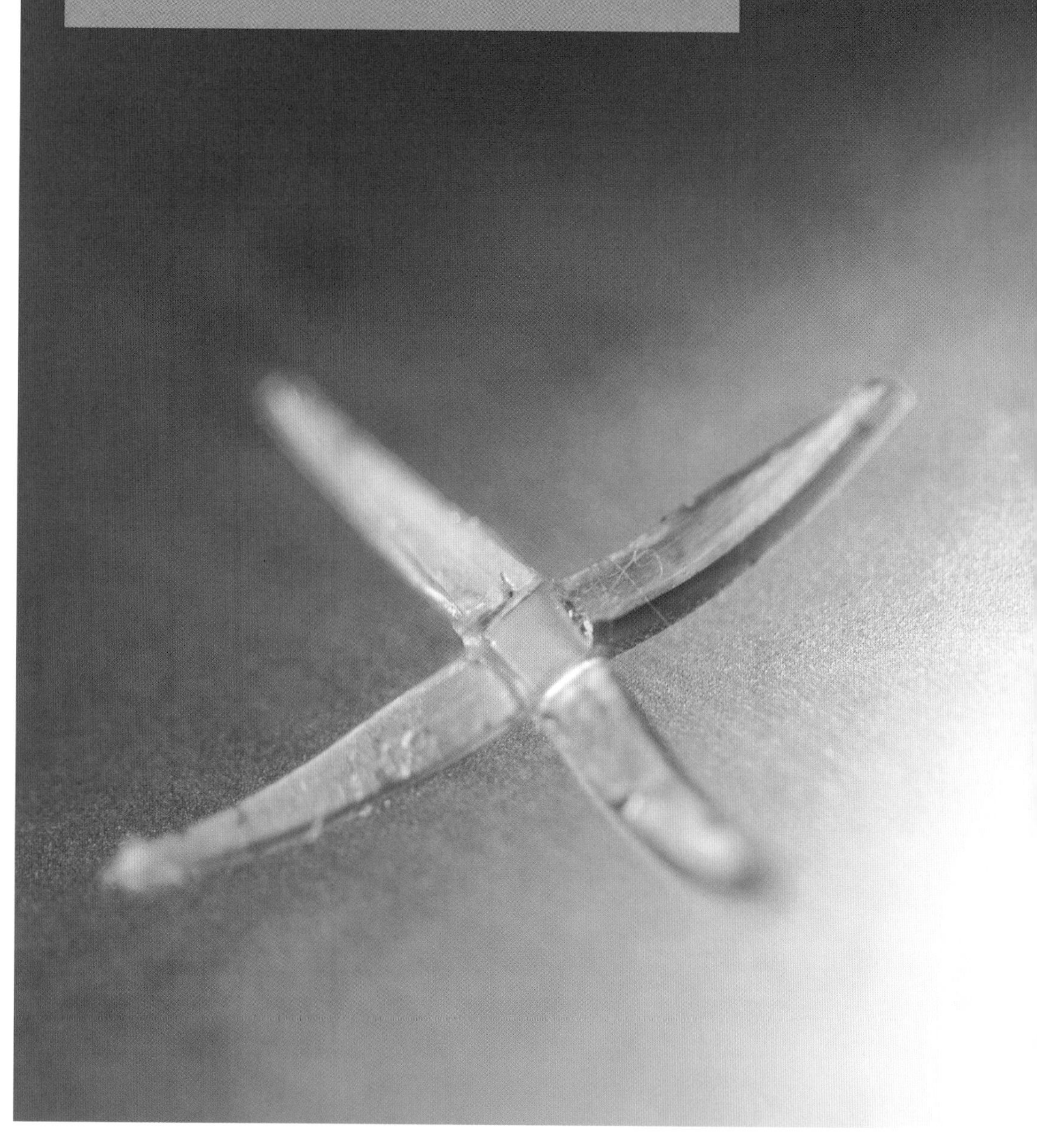

Liquid crystal elastomers (LCEs) are shape-responsive materials that can react to stimuli in the environment. These materials are biocompatible and can be used as scaffolds for tissue engineering that can actively direct cellular growth and differentiation. LCEs also exhibit interesting structural properties and have been shown to increase in stiffness when subjected to compression. They can thus be employed as structural materials that withstand repeated compression or record a history of compression and tension in a product-testing scenario.

I
II
III
IV
V

x axis *z* axis *y* axis

CONTENTS

50 percent polysiloxane, 40 percent liquid crystal, 10 percent cross-linker

APPLICATIONS

Sensing, structural applications, tissue engineering

TYPES / SIZES

Vary

TESTS / EXAMINATIONS

Laboratory tests

LIMITATIONS

For interior use only

FUTURE IMPACT

Advanced polymers for use in biomedical implants and tissue engineering, products that emulate organic structures and tissues

COMMERCIAL READINESS

CONTACT

Rice University
Verduzco Laboratory
6100 Main Street, MS-362, Houston, TX 77005
713-348-6492
www.polymers.rice.edu
rafaelv@rice.edu

LIGHT-EMITTING BIOPOLYMER

Photoluminescent Bioplastic

MATERIAL

Bioplastics offer a broad range of characteristics suitable for design and architectural applications. Architect Peter Yeadon experiments with various configurations of bioplastics and dopes them with selected impurities to enhance their properties. Yeadon eschews conventional biodegradable polymers like polylactic acid (PLA), preferring feedstock derived from industrial biowaste that can withstand deterioration from solvents, such as tetrahydrofuran and acetone. He creates photoluminescent bioplastics out of triglycerides from biodiesel production, collagen from cattle tissue, and light-emitting colloids.

In 2015 his firm, Yeadon Space Agency, collaborated with industrial designer Danielle Storm Hoogland to produce an outdoor furniture prototype called the BioBench. The bench absorbs sunlight throughout the day and illuminates at night for enhanced visibility.

CONTENTS
Triglycerides, collagen proteins, deionized water, europium aluminate

APPLICATIONS
Luminous objects and surfaces

TYPES / SIZES
Vary

ENVIRONMENTAL
Passive illumination, potential to use biowaste as source material

LIMITATIONS
Temporary illumination

FUTURE IMPACT
Replacement of photoluminescent petroleum-based plastics with high-performance, bio-based alternatives; new material stream for biowaste

COMMERCIAL READINESS

CONTACT
Yeadon Space Agency
33 Flatbush Avenue, 6th Floor, Brooklyn, NY 11217
www.yeadonspaceagency.com
agent@yeadonspaceagency.com

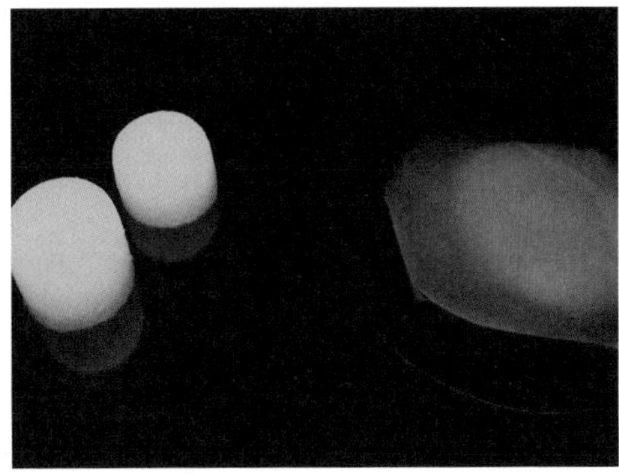

RECYCLED HDPE VESSELS

Plag

Plag represents designer Waël Seaiby's noble effort to upcycle the disposable plastic bag. Approximately one million such bags are used each minute across the globe, and over 90 percent wind up in landfills. Most are made of high-density polyethylene (HDPE), a polymer that can be used for much more demanding and enduring applications. Plag is a collection of upcycled, handcrafted vessels that evoke ceramic and glass craftsmanship. Because they consist of recycled shopping bags, Plag containers bear the bright pigments of their original HDPE feedstock, and exhibit whimsically distorted forms and rough edges in the spirit of their first use. In this way, the Plag collection transforms one of the least valuable uses of plastic into a high-quality application.

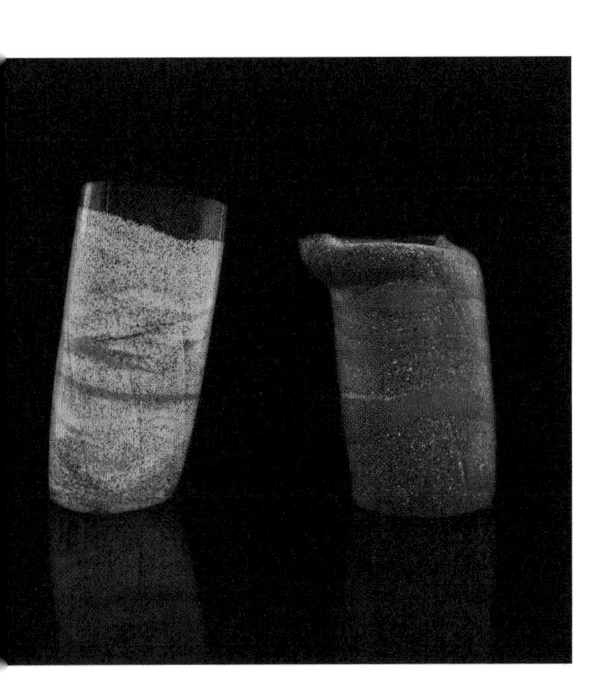

CONTENTS
Recycled HDPE

APPLICATIONS
Tableware, vases, decorative objects

TYPES / SIZES
Custom colors and finishes, based on original HDPE materials

ENVIRONMENTAL
Upcycled waste plastic, 100 percent recyclable material composition

LIMITATIONS
Requires custom arrangements with local recyclers, not dishwasher safe

FUTURE IMPACT
Reduced waste burden in landfills, upcycled objects should far outlast the first life of the HDPE bags

COMMERCIAL READINESS

CONTACT
Waël Seaiby
Edinburgh, UK
+44 (0) 7447540788
www.waelseaiby.com
info@waelseaiby.com

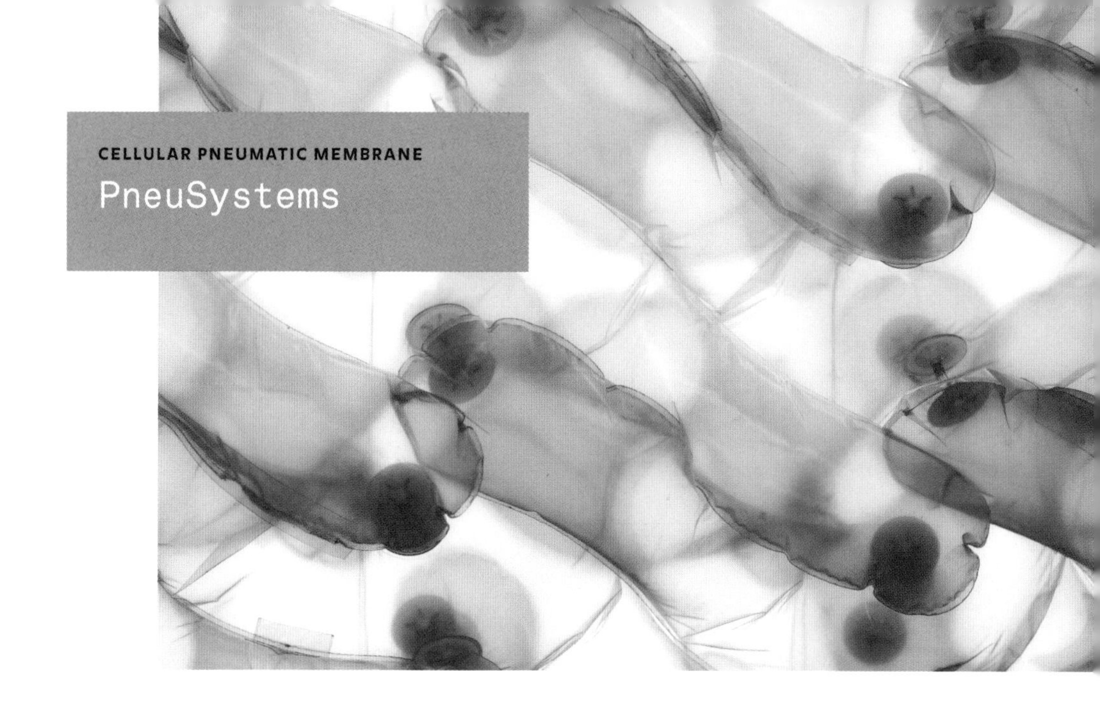

CELLULAR PNEUMATIC MEMBRANE

PneuSystems

PRODUCT

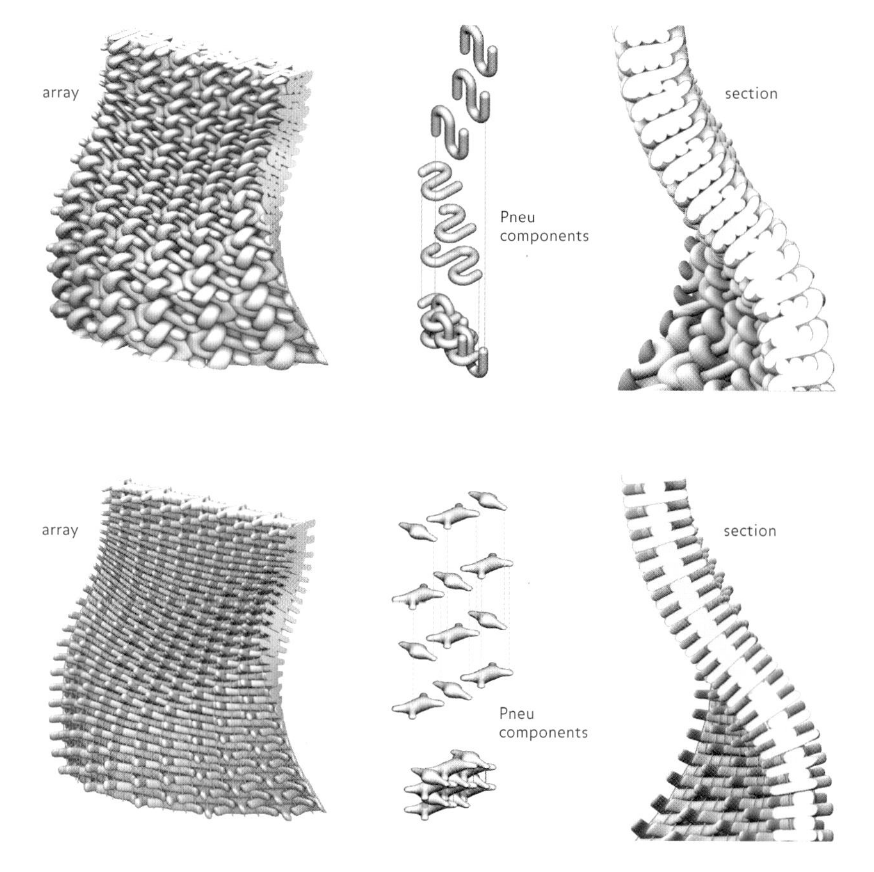

PneuSystems is a modular, air-supported surface designed to function as a lightweight, adaptable architectural skin. Developed by research and design firm RVTR, the project was inspired by membranes in natural organisms as well as the engineered systems research conducted at Stuttgart's Institute for Lightweight Structures and Conceptual Design. Three primary investigations informed the development of PneuSystems: aggregated structures formed by nested cellular assemblages, energy performance of air-entrained systems, and rapid digital prototyping processes employing custom tools and jigs. Unlike air-supported structures with a single inflated volume, PneuSystems exhibits systemic resilience through the redundancy of its many interlocking components.

CONTENTS
Polymer pillow modules, air

APPLICATIONS
Responsive building envelopes

TYPES / SIZES
Vary

ENVIRONMENTAL
Ultralight materials with insulating capacity

LIMITATIONS
Susceptible to damage from sharp objects

FUTURE IMPACT
Revitalization of interest in lightweight structures based on inflatable polymer components

COMMERCIAL READINESS

CONTACT
RVTR
305 West Liberty Street, Ann Arbor, MI 48103
734-834-9385
www.rvtr.com
geoff@rvtr.com

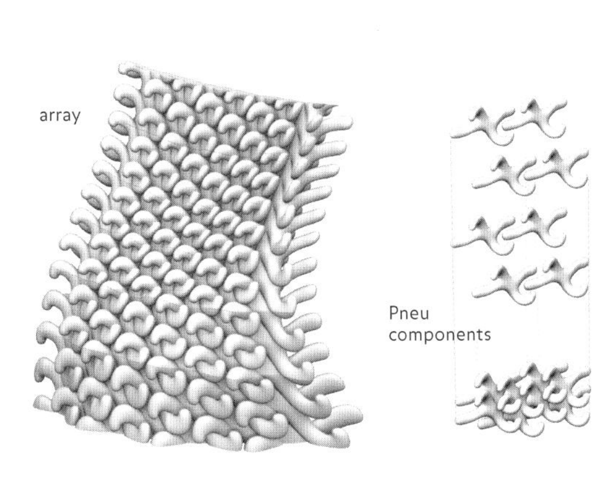

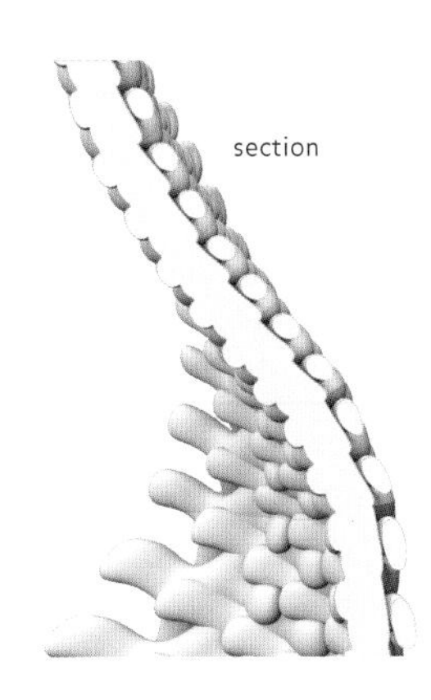

SELF-HEALING POLY(UREA-URETHANE) ELASTOMER

Self-Healing PUU

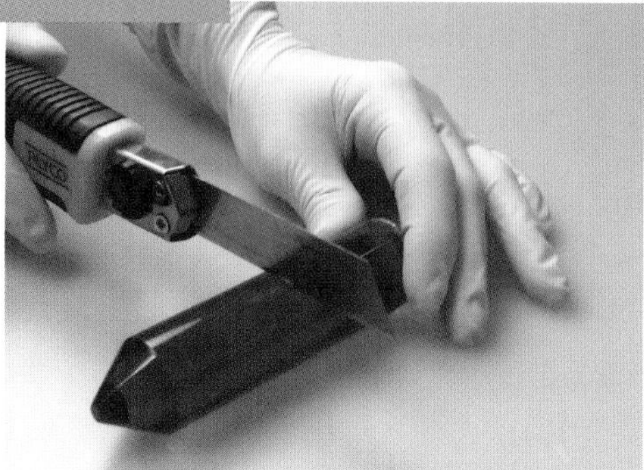

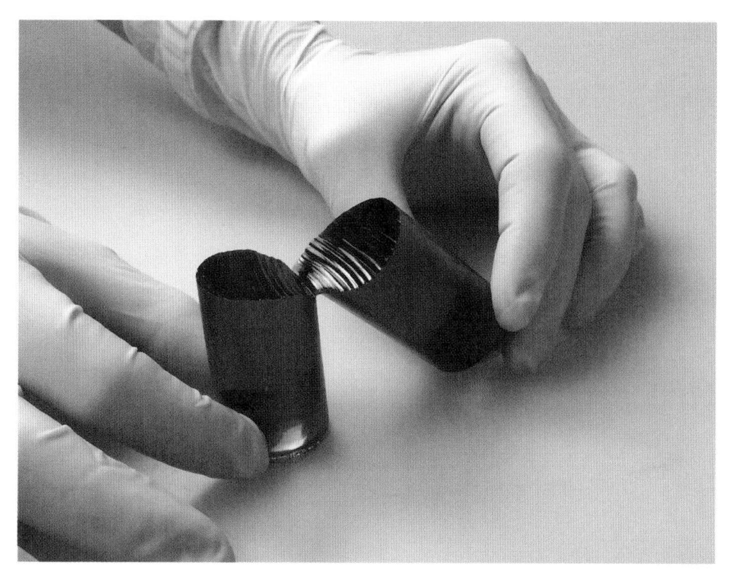

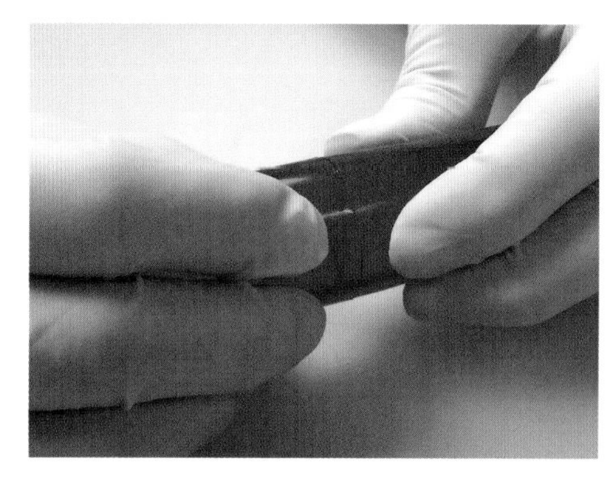

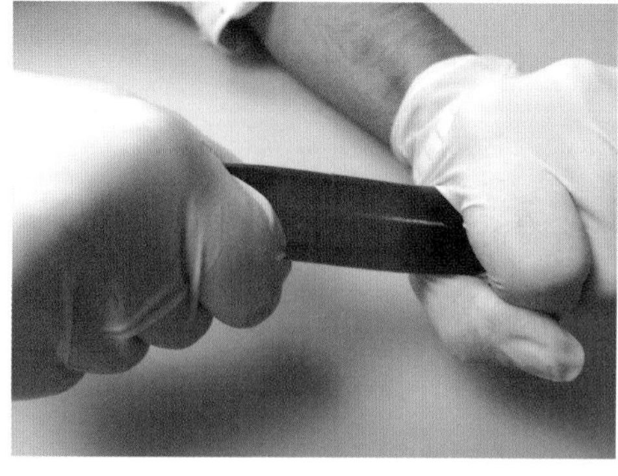

Self-healing plastics automatically restore themselves when damaged. These materials conventionally mend broken bonds with the aid of an external stimulus, such as light, heat, or changing pH level. Scientists at Spain's IK4-CIDETEC Research center have created a material capable of spontaneous quantitative healing without an external stimulus or catalyst—something previously considered impossible. The trick is the use of aromatic disulfides that undergo a chemical reaction in which they exchange bonds at room temperature, otherwise known as a metathesis reaction. As a result, the self-healing elastomer can not only repair small cracks but also mend itself after being cut entirely in half. This is the first self-healing material technology with this extraordinary and advantageous capacity.

CONTENTS
100 percent poly(urea-urethane)

APPLICATIONS
Adhesives, sealants, dynamic seals, coatings, dampening materials, machine components, vehicles, bridges, railway elastomer mats

TYPES / SIZES
Vary

ENVIRONMENTAL
Increased material longevity, delayed technical obsolescence

TESTS / EXAMINATIONS
ISO 37, ISO 815

LIMITATIONS
Maximum temperature of use 392 °F (200 °C)

FUTURE IMPACT
Increased polymer durability and reduced frequency of required maintenance

COMMERCIAL READINESS
● ○ ○ ○ ○

CONTACT
IK4-CIDETEC Research Center
Parque Científico y Tecnológico de Gipuzkoa, Paseo Miramón, 196, Donostia, San Sebastián, Gipuzkoa 20009, Spain
+34 943309022
www.cidetec.es
cidetec@cidetec.es

VARIABLE VACUUM-FORMED PLASTIC

VarVac

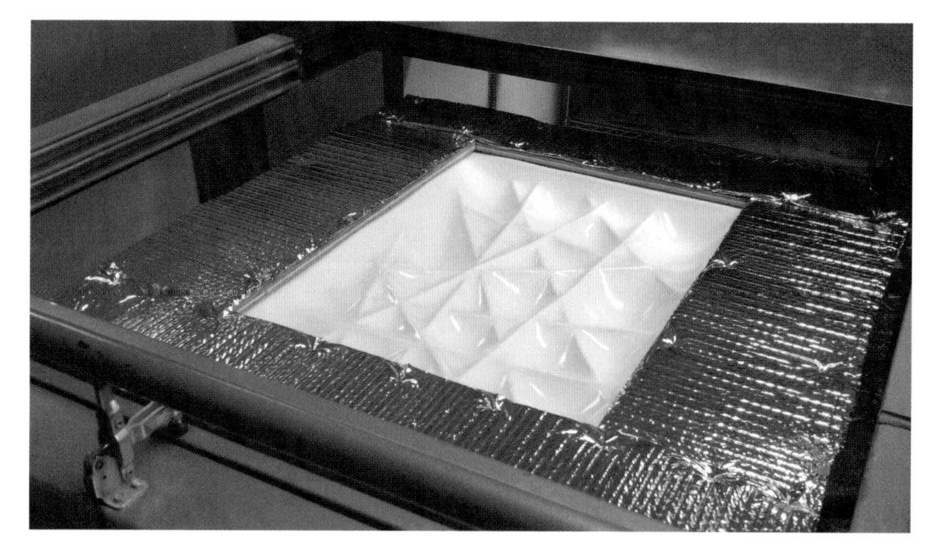

PROCESS

Although vacuum-forming molds are typically costly, the parts they produce are relatively inexpensive due to the economy of mass production. Vacuum-formed thermoplastics display several other positive attributes. For example, the plastics are light and, when formed correctly, require little structural support. They are durable, impervious to water, and noncorrosive. Most thermoplastics are recyclable, and new bioplastics are produced from renewable resources.

Minneapolis- and Vancouver-based design and research firm HouMinn Practice explores vacuum forming as a method for producing low-cost, multifaceted architectural surfaces, despite the fact that the cost of the mold and its inflexibility typically prohibit aggregation with difference. The VarVac system is made by suspending heated plastic sheets in midair over a wood frame with connective wires. The end of each wire may be manually shifted along the *x* or *y* axis, providing the opportunity for endless variability via simple, rapid changes during setup. Once heated, depressions form on the plastic sheet in the voids between the wires. The cooled material is then cut with a CNC router to form voids, which may be strategically located for functions such as acoustic absorption or access to light switches. When panels are combined to form a unified surface, the result is a heterogeneous field that can be tuned precisely to meet local needs and preexisting physical conditions.

CONTENTS

PETG, PS, or other thermoplastics

APPLICATIONS

Wall and ceiling finishes, acoustically absorptive surfaces

TYPES / SIZES

Vary

TESTS / EXAMINATIONS

Permanent installation in the departmental office of the University of Minnesota School of Architecture

LIMITATIONS

Size limited to the variable cable mold frame dimensions

FUTURE IMPACT

Opportunities for low-cost mass customization: fabrication of variegated, nonrepeating surfaces based on a single mold

COMMERCIAL READINESS

● ○ ○ ○ ○

CONTACT

HouMinn Practice
6344 Warren Avenue South, Edina, MN 55439
612-669-2603
www.houminn.com
marc@houminn.com

CELLULOSE-BASED POLYMER

Zeoform

One of the first plastics was made from cellulose. In 1855 scientist Alexander Parkes created the first thermoplastic out of nitrocellulose and camphor. Later named "celluloid" in 1872, the material became the basis for all photography and motion picture films—until its replacement with cellulose acetate in the twentieth century. Despite this natural origin for plastic, cellulose is rarely found in modern polymers, as they are primarily derived from fossil fuels.

Cellulose is a critical structural material found in the cell walls of green plants and is the most abundant organic polymer on earth. Depending on the plant species, cellulose content can range between 40 and 90 percent. Australia-based company Zeoform developed a method to replicate the natural process of hydroxyl bonding, which provides strength to collective cellulose fibers in plants. The company makes its new polymer from cellulose and water. It extracts pure cellulose from reclaimed and recycled sources, such as paper, hemp, and natural fabrics, and its fabrication process requires no glues, binders, or additives. According to Zeoform, the new material is a universal feedstock that can be sprayed, molded, or formed into various shapes, and can substitute for most plastics, woods, and composite materials used today.

CONTENTS

Cellulose by-products, such as hemp, agricultural biomass, recycled paper, cotton, rice, jute, cane, wood, or bamboo; water

APPLICATIONS

Facade cladding, doors, millwork, countertops, tiles, structural shapes, handrails, furniture, handles, switches, light fittings, decorative vessels, surfboards, musical instruments

TYPES / SIZES

Offered in densities ranging from that of Styrofoam to ebony; various natural pigments, coatings, and textures available

ENVIRONMENTAL

100 percent nontoxic, biodegradable, and compostable; repurposes agricultural, industrial, or consumer waste products; sequesters stored CO_2

TESTS / EXAMINATIONS

Environmental impact, life cycle tests

LIMITATIONS

Precision components not yet achievable at an industrial scale, cannot yet be 3D printed

FUTURE IMPACT

Potential to replace plastic and wood composites in a wide variety of applications with a significantly more environmentally responsible feedstock

COMMERCIAL READINESS

●●●●●

CONTACT

Zeoform
PO Box 324, Mullumbimby,
New South Wales 2482, Australia
+612 66844553
www.zeoform.com
info@zeoform.com

6
Glass

> Captured Light

> German author Paul Scheerbart, best known for his 1914 book *Glasarchitektur*, is often cited as a significant influence on the advancement of modern architecture. His visions of a crystalline world of soaring, transparent glass edifices captivated the imaginations of architects such as Bruno Taut, who dedicated his Glass Pavilion at the Werkbund exhibition (1914) to Scheerbart. The International Style that emerged in the 1920s elevated the role of glass in architecture, expanding its application from traditional punched openings into large vertical expanses offering panoramic views.

Although some historians would interpret the ensuing decades of glass-dominated building as a fulfillment of Scheerbart's aspirations, the reality is only a partial consummation.

Today glass skins have largely replaced the brick facades of early twentieth-century cities. Yet, with its emphasis on absolute flatness and optical clarity, modern float glass is inherently designed to disappear—to be a nonmaterial. In contrast, Scheerbart imagined glass as an active substance, rich in color and geometry, and inspired by living organisms. At the culmination of Scheerbart's novel *The Gray Cloth and Ten Percent White: A Ladies Novel* (1914), his protagonist, architect Edgar Krug, gazes up into the dome of his glass tower and proclaims: "Dragonfly wings!... Birds of paradise, fireflies, lightfish, orchids, muscles, pearls, diamonds, and so on, and so on—All that is beautiful on the face of the earth. And we find it all again in glass architecture. It is the culmination—the cultural peak!" [1]

Although historians have long referenced this kind of imaginative vision as evidence of the fantastic nature of Scheerbart's utopian ideals, it is increasingly becoming reality. Today, a century after Scheerbart's death, we see a burgeoning manifestation of his original kaleidoscopic ambitions, which he described as "glass elements in every possible color and form as a wall material." [2] A survey of next-generation glass technologies suggests that this reassertion of glass as a substantive material is intimately connected with new glass functionalities and processes. New glass printing methods, high-performance properties, biologically inspired behaviors, and renewable energy applications privilege capacities that impart shape, density, pattern, and character to the material, empowering it to transcend its role as an evanescent surface.

In 2011 German designer Markus Kayser developed a mobile, solar-powered machine for printing glass with sand. Kayser transported his Solar Sinter into the Sahara Desert for a number of experiments, and he created a series of small vessels via this sun-powered selective laser sintering process. The resulting objects are simultaneously rugged and precise, composed of opaque strata with a densely encrusted surface. ← **Kayser later joined the Mediated Matter Group at MIT, where he and a team of researchers developed the G3DP process, an additive manufacturing method that produces optically transparent glass objects (see page 178).** The innovative process employs dual heated chambers for melting and annealing, and the objects form as the molten glass is deposited via an alumina-zircon-silica nozzle.

Advances in glass processing have also permitted the realization of unprecedented capabilities. Scientists at McGill University in Montreal studied nacre and

other robust natural materials and found that they are composed of microstructural blocks separated by weak boundaries. **↑ Comprehending this structure as a way to prevent crack propagation, the researchers used a laser to engrave networks of microfissures in borosilicate glass. The resulting material exhibits two hundred times greater toughness than conventional glass (see page 174).** In 2011 Corning scientist Terry Ott developed a method to mass-produce Willow, a glass with the thickness of paper.[3] Created for thin displays and portable devices, Willow is made using a continuous roll-to-roll process that produces flexible strips three football fields long.

One of the swiftest-growing uses of glass is glazing for energy-related applications. Manufacturers are keen to mass-produce optically transparent windows that generate electricity, given the potential market for energy-based retrofits, particularly in buildings with ample surface area. SolarWindow Technologies offers a version that consists of transparent organic photovoltaic coatings applied to a glass substrate (see page 188). **↗ Idaho-based Solar Roadways makes high-strength glass modules that encapsulate photovoltaics, sensors, and LEDs for use in constructing intelligent highways (see page 184). The interlocking hexagonal pavers**

provide power for road illumination, heating elements, and electric car battery charging. Arup's SolarLeaf glazing system incorporates living algae to harvest solar energy by growing biomass (see page 186). Each glass module is essentially a bioreactor that automatically provides more shade where needed. Excitable glass technology may also be used for solar thermal control. Researchers at the National Institute of Advanced Industrial Science and Technology (AIST) in Japan have developed an electrochromic mirror window that switches between transparent and mirror states with the flip of a switch (see page 190). Designed for solar envelope applications, the switchable mirror is estimated to reduce energy consumption in buildings and vehicles by over 30 percent.

Researchers also see glazing as a promising territory for emulating biological systems. **↑ MIT scientists have developed a responsive membrane window, called a Vascular Skin, which employs capillary networks of hydrogels (see page 192).** The hydrogels initiate a phase transition just above room temperature, modifying optical transmission in an environmentally responsive way. Meanwhile, researchers at the University of Toronto and Harvard have created an adaptive microfluidic glazing system that can moderate surface temperature like animal skin.[4] By circulating water through ultrathin channels similar to blood vessels, the window facilitates heat transfer when hot.

Bioresponsive, solar-activated, and ultra-high-performance glazings indicate the rise of a new generation of glass technologies. These materials depart from the long trajectory of optical clarity and physical homogeneity, which privilege the construction of expansive and monotonous enclosures, and instead prioritize enhanced functions and characteristics. Next-generation glass exhibits an unprecedented capacity to manipulate light—whether for energy harvesting, view modification, illumination, or thermal conditioning—and this ability will influence the way society interacts with glazed surfaces. In the words of the prophetic Scheerbart, glass architecture "lets in the light of the sun, the moon, and the stars, not merely through a few windows, but through every possible wall. . . . The new environment, which we thus create, must bring us a new culture."[5]

1 Paul Scheerbart, *The Gray Cloth: Paul Scheerbart's Novel on Glass Architecture*, ed. and trans. John A. Stuart (Cambridge, MA: MIT Press, 2001), 123.

2 Paul Scheerbart, *Glass Architecture*, ed. Dennis Sharp (Santa Barbara, CA: Praeger, 1972), 50.

3 See accessed January 17, 2016, https://www.corning.com/in/en/products/display-glass/products/corning-willow-glass.html.

4 See accessed January 17, 2016, http://wyss.harvard.edu/viewpressrelease/119/.

5 Scheerbart, *Glass Architecture*, 41.

STRATIFIED GLASS LIGHT WALL

Bar Code

The Bar Code project celebrates the vibrant and kaleidoscopic interplay between material and media. Designed by Höweler and Yoon Architecture, Bar Code is a site-specific, computer-controlled installation composed of stacked, backlit, low-iron glass. The CNC-water-jet-cut contour of the back edges produces an interior topography that is visible through the surface and glows when lit from the sides. The curved glass mass acts like a large lens, focusing and dissipating light based on its geometry. The low-iron glass is optically clearer than regular soda-lime float glass and allows for better light transmission and more accurate color rendition.

Höweler and Yoon designed the contoured back surface to form a continuous topography of concave and convex curves. To eliminate material waste, the designers calibrated the curves to use both sides of a single cut. The edge lighting occurs through an array of custom RGB LED elements mounted on a sliding armature. The LED matrix enables the mixing of colors and the individual control of each source. With this precisely controllable lighting array, users can generate a low-resolution image and create a catalog of behaviors based on color, light intensity, and speed. The result is an illuminated screen that utilizes the material depth and light-channeling properties of glass to create volumetric light, as opposed to a conventional surface-based display or direct light source.

CONTENTS

Contoured low-iron glass, multicolored LED matrix

APPLICATIONS

Feature walls, lighting, visual communication, public art

TYPES / SIZES

Vary

ENVIRONMENTAL

Low-energy light source

TESTS / EXAMINATIONS

Permanent installation at Carr Realty in Washington, DC

LIMITATIONS

Structural accommodations must be made for the cumulative weight of stacked glass

FUTURE IMPACT

Explorations of intersections between digitally controlled illumination and light-propagating materials

COMMERCIAL READINESS

CONTACT

Höweler and Yoon Architecture LLP
150 Lincoln Street, 3A, Boston, MA 02111
617-517-4101
www.hyarchitecture.com
info@hyarchitecture.com

FRACTURE-RESISTANT GLASS

Bendable Glass

At McGill University's Laboratory for Advanced Materials and Bioinspiration, Professor Francois Barthelat researches natural structures such as nacre and teeth to understand their inherent mechanical properties. Like many material scientists, Barthelat is intrigued by the notion that biology holds the answers to our technological ambitions for more resilient materials.

Barthelat and his team discovered that nacre consists of layers of microstructural blocks that are loosely held together by weak boundaries. The scientists sought to model these microfissures in glass to test its mechanical performance and used a laser to engrave networks of three-dimensional cracks in borosilicate glass slides. They also filled the gaps between the jigsaw-like patterns with polyurethane, intended to function as a kind of cement. The resulting sample far outperformed standard glass, with over two hundred times the toughness of the untreated material. This durability imparts the new glass with more flexibility than the typical version, which is brittle. The secret lies in the strategic use of weak boundaries—as seen in expansion joints used in buildings—to diminish crack propagation and material failure. This successful approach represents a pathway for toughening a variety of ceramic materials based on similar natural principles.

CONTENTS
Laser-engraved glass, polyurethane

APPLICATIONS
Shatter-resistant glass windows, screens, and objects

TYPES / SIZES
Initial sample size 1 9/16 × 7/8" (40 × 22 mm)

ENVIRONMENTAL
Increases material longevity

TESTS / EXAMINATIONS
Laboratory tests

FUTURE IMPACT
Development of tougher, more resilient materials using the nature-inspired strategy of weak boundaries to counter crack propagation

COMMERCIAL READINESS
●○○○○

CONTACT
McGill University
Department of Mechanical Engineering
817 Sherbrooke Street West, Montreal, Quebec, Canada H3A 2K6
514-398-6318
www.mcgill.ca/mecheng/
reception.mechanical@mcgill.ca

CELLULAR GLASS THERMAL INSULATION

Foamglas

Thermal weak spots in building facades, known as thermal bridges, often thwart the goal of more energy-efficient architecture. Buildings commonly exhibit linear thermal bridges at key structural connections, such as at the foundation, floor lines, or parapet. Structural materials like masonry, steel, and concrete are relatively poor insulators, yet they occupy these critical junctures for load-bearing purposes.

Foamglas is a cellular glass insulation developed for such load-bearing applications. A dimensionally stable material with a high compressive strength, it is noncombustible, waterproof, impervious to water vapor, pest resistant, and acid resistant. Foamglas slabs, boards, and tapered wedges are designed to occupy critical structural zones in building envelopes that are typically thermal weak spots. Foamglas Perinsul HL (high load), for example, has a compressive strength of 400 psi and a thermal conductivity of R-2.5 per inch of thickness. Combining excellent structural and insulation properties, the material is appropriate for use in highly thermally efficient buildings, such as Passive House projects.

CONTENTS

60 percent or more recycled glass plus feldspar, sand, soda, ferric oxide, sodium sulfate, and carbon black

APPLICATIONS

Insulation for building envelopes, flat or tapered roofs, and below-grade applications

TYPES / SIZES

Available in slabs and boards with a density range of 100 to 200 kg/m³, thermal conductivity from □ D ≤ 0.036 W/(mK); typical slab dimension 23 5/8 × 17 11/16 (600 × 450 mm), thickness range 1 9/16"–7 7/8" (40–200 mm)

ENVIRONMENTAL

Superior thermal performance and solution for structural thermal bridging; nontoxic and low-emitting material with zero ozone-depleting substances, flame retardants, VOCs, binding agents, mutagens, or carcinogens

TESTS / EXAMINATIONS

CEN Keymark, Passive House construction certification

FUTURE IMPACT

Potential to elevate the thermal performance of new construction dramatically; with further technical development, possibility to replace masonry altogether as a load-bearing material

COMMERCIAL READINESS

● ● ● ● ●

CONTACT

Pittsburgh Corning Europe
Albertkade 1, Tessenderlo 3980, Belgium
+32 (0) 13661721
www.foamglas.com
info@foamglas.com

3D-PRINTED OPTICALLY TRANSPARENT GLASS

G3DP

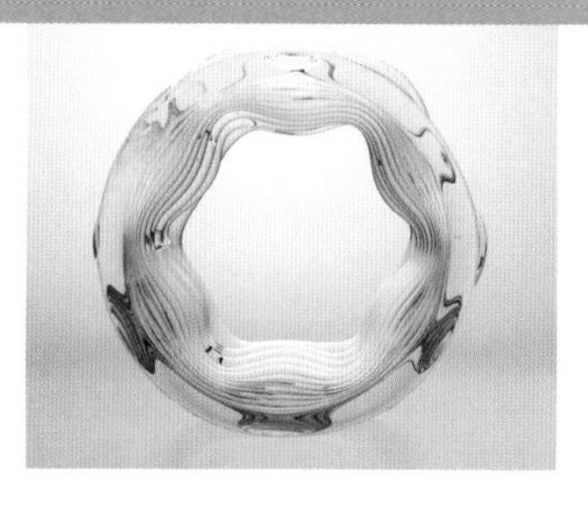

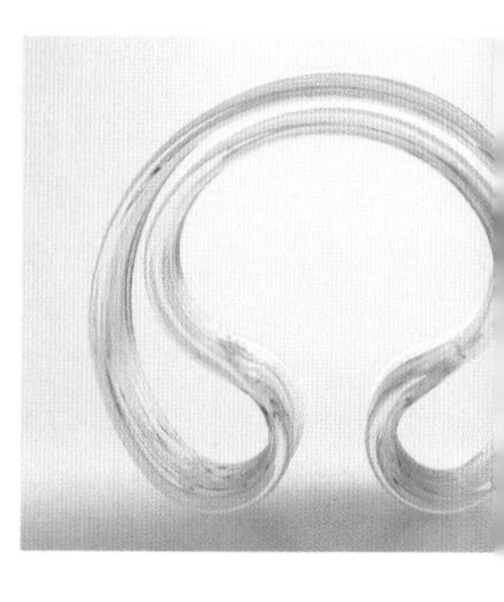

PROCESS

From the discovery of glass bead-making processes in ancient Egypt to the development of modern industrial flat glass production, glass has experienced fundamental changes in its manufacture over the centuries. G3DP represents yet another chapter in glass making.

G3DP is a digital fabrication method to print glass with optical transparency. The process allows for a high degree of control over geometric and optical variations, which in turn influence the form, transparency, color, reflection, and refraction of printed glass. The additive manufacturing method employs an alumina-zircon-silica nozzle and two heated chambers. The upper chamber functions as a kiln while the lower one anneals the glass objects. The upper chamber has an operating temperature of 1,900 °F (1,038 °C) and is of sufficient size to produce a complete object that is printed in a single pass.

Unveiled at the Cooper Hewitt, Smithsonian Design Museum in 2016, G3DP represents the synthesis of modern technologies with ancient glass techniques and generates novel glass structures with numerous potential applications.

CONTENTS
Soda-lime glass

APPLICATIONS
Transparent glass objects

TYPES / SIZES
Vary

ENVIRONMENTAL
Recyclable

LIMITATIONS
Requires maintaining a constant high temperature

FUTURE IMPACT
Expanded study of geometric complexity and optical phenomena in transparent glass

COMMERCIAL READINESS
● ● ● ● ●

CONTACT
Mediated Matter Group
Massachusetts Institute of Technology Media Lab
75 Amherst Street, Room E14–433B, Cambridge, MA 02139-4307
617-324-3626
matter.media.mit.edu
ked03@media.mit.edu

MOISTURE-ABSORBING PLASTER INCLUSION

Porous Glass

Porous Glass is a flake-based glass material for incorporation in plaster. Made of Vycor glass, which is 95 percent silica, Porous Glass dramatically enhances plaster's ability to absorb, store, and release moisture. This capacity is increasingly valuable in the context of tighter building insulation standards and construction methods that can lead to increased levels of humidity in interior spaces. Persistent moisture in buildings is the leading cause of mold and mildew, which can result in sick building syndrome for building occupants. Porous Glass outperforms conventional humidity-regulating products, such as fiberboard or zeolite, as the inorganic material can absorb and release more moisture over both diurnal and seasonal cycles. The plaster additive is also nontoxic, noncombustible, and inexpensive.

CONTENTS
High-silica glass flakes, plaster

APPLICATIONS
Moisture-absorbing plaster

TYPES / SIZES
Vycor high-silica glass

ENVIRONMENTAL
Nontoxic, absorbs mildew-causing humidity, helps regulate indoor climate

TESTS / EXAMINATIONS
Adsorption and desorption tests

LIMITATIONS
Moisture-regulating effects may be hindered by additional layers of materials, such as wallpaper or paint

FUTURE IMPACT
Small yet effective material enhancements that make considerable improvements in indoor environmental quality

COMMERCIAL READINESS

CONTACT
Fraunhofer Institute for Silicate Research ISC
Neunerplatz 2, Würzburg 97082, Germany
+49 9314100229
www.isc.fraunhofer.de
info@isc.fraunhofer.de

SOLAR THERMAL GLASS AND WOOD CLADDING SYSTEM

Solar Activated Facade

Conventional facades attempt to decouple the building interior from the exterior. The Solar Activated Facade (SAF) is designed to bridge inside and outside via a thermally activated buffer. Originally invented by a Swiss architect in 1998, the SAF consists of slanted, horizontal slats of solid wood combined with a back-vented glass curtain wall. The exterior glass protects the wood from the elements and preserves the thermal buffering effect, trapping heat inside the envelope for delayed release. Behind the SAF is a conventional structural wall of insulated wood or masonry construction.

In the winter solar radiation is transmitted through the glass curtain wall and absorbed by the wood louvers. The wood gradually warms during the daylight hours and releases heat after sunset over the course of four to twelve hours. During this period, building heat losses are minimized, as the system functions as a thermal barrier against the cold. In the summer the sun strikes the wooden louvers at a steeper angle and the slats are self-shadowing. As a result, the facade absorbs less solar radiation than in the winter.

The Solar Activated Facade was first offered in Switzerland under the name Lucido, and Nelson Solar GmbH has more recently introduced SAF to the American market.

CONTENTS
Glass, wood, aluminum framing

APPLICATIONS
Building envelopes, thermal mass walls

TYPES / SIZES
Maximum width or height of 8' (2.44 m)

ENVIRONMENTAL
Heat storage and delayed release via thermal mass, high insulating capacity, low embodied energy

TESTS / EXAMINATIONS
Tests conducted at the Swiss Federal Laboratories for Materials Science and Technology, Dübendorf, Switzerland

LIMITATIONS
Requires solar exposure for optimal performance

FUTURE IMPACT
Increased energy savings, particularly for buildings in northern climates

COMMERCIAL READINESS

CONTACT
Nelson Solar
Rigistrasse 33, Cham ZG 6330, Switzerland
+41 (0) 415581986
www.nelson-solar.com
eric.nelson@nelson-solar.com

ELECTRICITY-GENERATING SOLAR ROAD MODULES

Solar Roadways

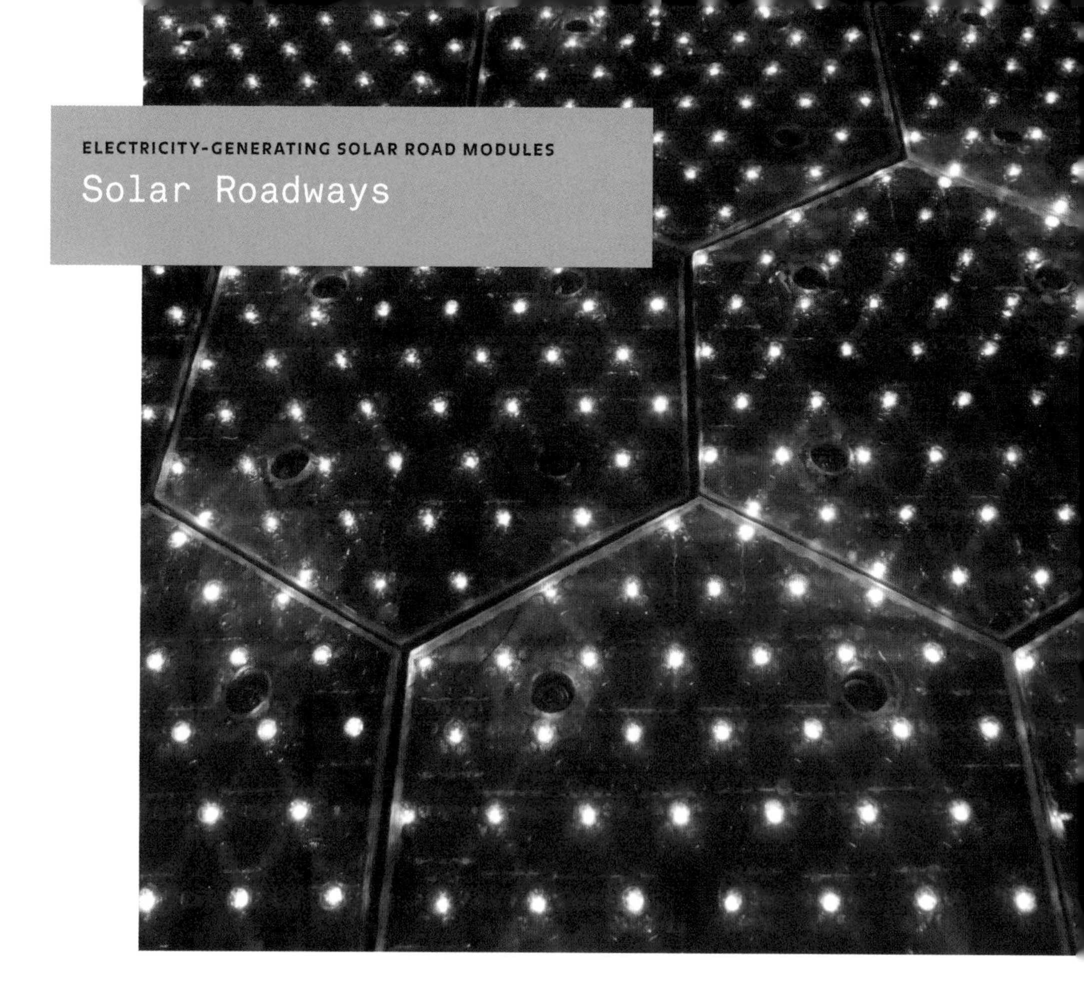

To be most effective, solar power requires an expansive area for energy harvesting as well as a limited distance across which it must convey the electricity produced, to avoid significant signal reduction. Solar farms produce large quantities of energy but typically exist in the hinterland due to space requirements—a location that necessitates long-distance travel for power delivery.

Developed by Idaho-based inventors Julie and Scott Brusaw, Solar Roadways combines solar power generation with road infrastructure. Based on their assessment that there are 27,800 mi² (72,000 km²) of concrete and asphalt paving in the United States alone, the Brusaws created a multifunctional paving module that—if used to replace traditional paving—would generate up to three times the country's current electricity demand.

The interlocking, hexagon-shaped pavers are surfaced with durable textured glass and contain layers of photovoltaics, sensors, and LEDs. Harvested energy may be used to power homes and businesses served by the roadways, illuminate lane lines and signage, and melt snow and ice on the road surface. The entrepreneurs are also developing ways to power electric cars through the road's surface, either via direct induction in self-charging wheels or with charging stations in parking lots.

CONTENTS

High-strength glass, photovoltaics, sensors, LEDs, heating elements

APPLICATIONS

Roads, bicycle paths, sidewalks, parking lots, solar-exposed structured parking; renewable power generation

TYPES / SIZES

Each hexagonal paver is 4.39 ft² (0.41 m²) and generates 48 watts of electricity

ENVIRONMENTAL

Net-zero electricity generation, reduced power necessary for road lighting, reduced material resources required for winter road maintenance

TESTS / EXAMINATIONS

Field testing under U.S. Department of Transportation Small Business Innovation Research (SBIR) program, Phase II SBIR, Phase IIB contracts

LIMITATIONS

Costlier than conventional roadbed construction

FUTURE IMPACT

Potential to combine renewable power generation and vehicular infrastructure with far-reaching environmental, economic, and safety benefits

COMMERCIAL READINESS

CONTACT

Solar Roadways
721 Pine Street, Sandpoint, ID 83864
208-946-2937
www.solarroadways.com
questions@solarroadways.com

BIOREACTOR FACADE SYSTEM

SolarLeaf

PRODUCT

SolarLeaf is the world's first bioreactive facade. Developed by Arup engineers in collaboration with the Strategic Science Consult of Germany and Colt International in a building designed by Splitterwerk for the 2013 International Building Exhibition in Hamburg, the algae-infused curtain wall system generates renewable power from algal biomass whose growth is encouraged by the sun. The system transports mature algae and heat within a closed loop to a building energy management center, which harvests the algal biomass via floatation and the heat via a heat exchanger. Because of its integration with building infrastructure, the energy management center utilizes any excess heat to provide hot water and heat to the building. Furthermore, the bioreactive facade functions as a dynamic shading device. The portions of the building exposed to the greatest intensity of sunlight experience the most algal growth—and hence, the maximum degree of shading.

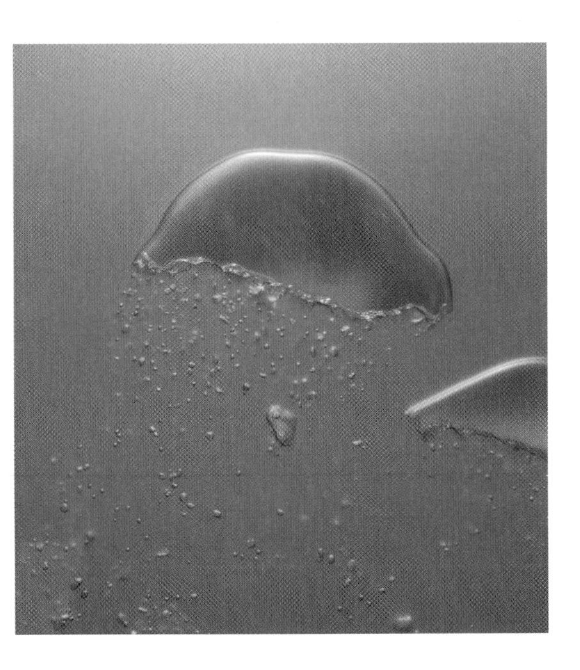

CONTENTS

Four layers of argon-insulated glass, water, algae culture, growth medium, plastic scrubbers, compressed air, inflow and outflow pipes

APPLICATIONS

Building envelopes, renewable energy generation, exterior shading

TYPES / SIZES

Standard bioreactor measures 8' 2" × 2' 4" (2.5 × 0.7 m)

ENVIRONMENTAL

Photobioreactor process supplies heat and hot water from the cultivation of biomass, automatic shading reduces building energy consumption

TESTS / EXAMINATIONS

In situ tests conducted at the BIQ house, Hamburg, 2013

LIMITATIONS

Maximum operating temperature of 104 °F (40 °C), requires routine inspection and maintenance

FUTURE IMPACT

Cultivation of living organisms for optimizing building operations processes, reduced reliance on fossil fuels, incorporation of energy-harvesting and mechanical systems as primary elements of the building facade

COMMERCIAL READINESS

CONTACT

Arup Deutschland GmbH
Jachimsthaler Strasse 41, Berlin 10623, Germany
+49 308859100
www.arup.com
berlin@arup.com

SOLAR ENERGY–HARVESTING WINDOW

SolarWindow

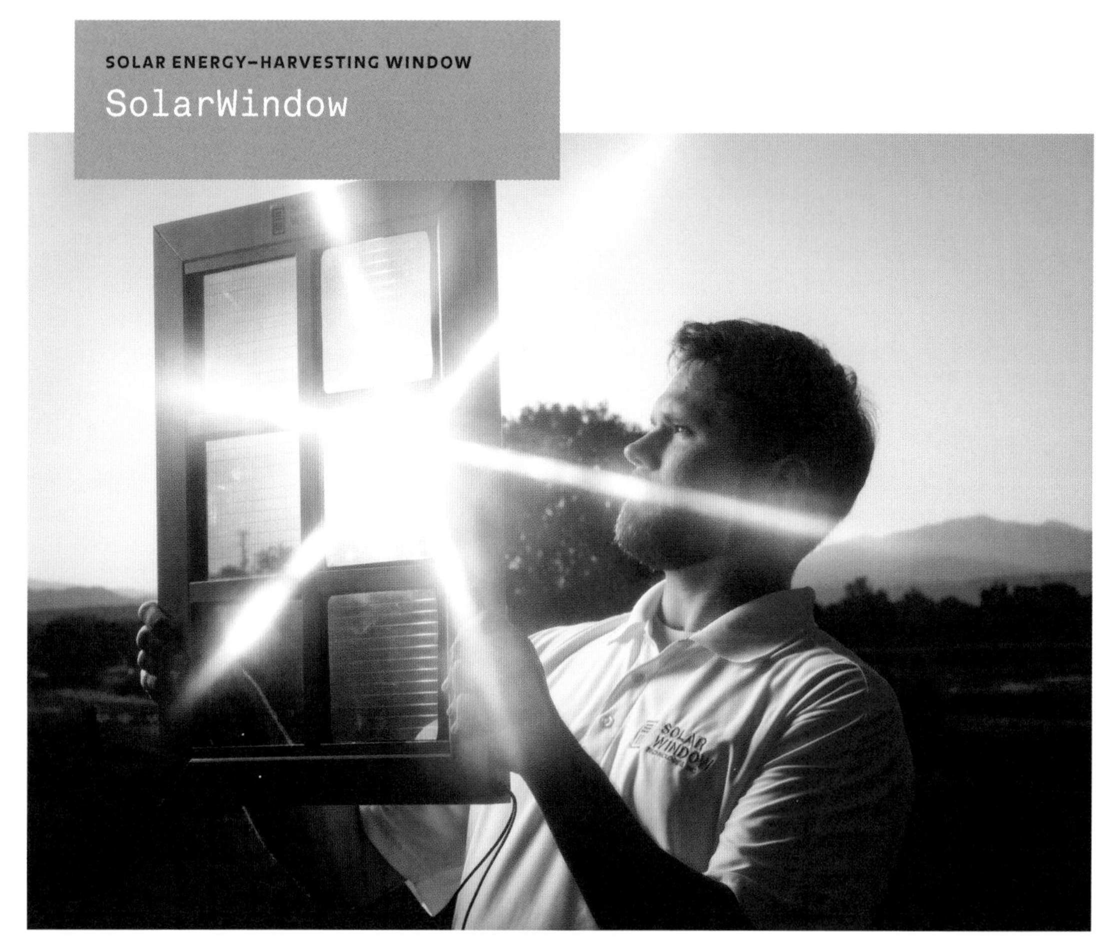

PRODUCT

Global energy demand continues to increase, and U.S. electricity demand alone is expected to grow 40 percent by 2032. According to the U.S. Department of Energy, commercial buildings consume more than one-third of this electricity. SolarWindow uses a transparent layer of organic photovoltaics (OPV) to harness energy at the location of building apertures. Designed especially for installation in high-rises, SolarWindow addresses the need to produce renewable energy for a structure with limited rooftop space. Unlike conventional photovoltaic systems, SolarWindow coatings may be applied to all sides of buildings, as they generate electricity in both natural and artificial light, including shaded, diffuse, and reflected lighting conditions.

The company's proprietary power production and financial model uses photovoltaic modeling calculations to evaluate and estimate renewable energy for a photovoltaic project. The power and financial model estimator takes into consideration geographic location, solar radiation for flat-plate collectors, climate zone energy use, and generalized building characteristics when estimating photovoltaic energy production and CO_2 equivalents.

CONTENTS

Transparent organic OPV coatings, glass (or plastic) substrate

APPLICATIONS

Renewable electricity generation

TYPES / SIZES

Vary

ENVIRONMENTAL

Solar power harvesting for all sides of a building, in shade as well as direct sunlight

TESTS / EXAMINATIONS

IEC photovoltaic standards, UL 1703, AC 365, ASTM IGU performance standards, FR 706, California Energy Commission, Florida State Energy Center

LIMITATIONS

OPV material thickness ranges from 500 to 800 nanometers

FUTURE IMPACT

Transformation of tall structures and buildings with significant window areas into solar power plants, potential to turn glazing retrofit projects into energy-saving strategies

COMMERCIAL READINESS

CONTACT

**SolarWindow Technologies Inc.
10632 Little Patuxent Parkway, Suite 406,
Columbia, MD 21044
800-213-0689
www.solarwindow.com
info@solarwindow.com**

ELECTROCHROMIC MIRROR WINDOW

Switchable Mirror

Researchers at Japan's National Institute of Advanced Industrial Science and Technology have developed a mirror that quickly shifts between reflective and transparent states. The switchable mirror consists of a thin layer of magnesium-titanium alloy film encapsulated between two layers of glass. The scientists introduce a gas (either 1 percent hydrogen or 20 percent oxygen) to induce the window's switching behavior.

In the development of early prototypes, researcher Kazuki Yoshimura and his team found that a magnesium-nickel alloy imparted a yellow tinge when transparent. However, the new magnesium-titanium version lacks this undesirable coloration. With the largest prototype at 24 × 28" (60 × 70 cm) in size, AIST scientists anticipate the switchable mirror will reduce energy consumed by air conditioning systems in buildings and automobiles by over 30 percent, simply by reflecting rather than admitting solar heat.

CONTENTS

Glass, transparent electrodes, magnesium and titanium thin film, adhesive electrolyte, ion storage layer

APPLICATIONS

Energy-saving windows, switchable privacy screens, electronic and optical devices

TYPES / SIZES

Vary, largest prototype window 24 × 28" (60 × 70 cm)

ENVIRONMENTAL

Reduced solar heat gain during the day and heat loss during the night

TESTS / EXAMINATIONS

Light transmittance and reflectance tests

LIMITATIONS

Requires the application of low-voltage electricity

FUTURE IMPACT

Increased control over light and heat transmission through building apertures and vehicular windows

COMMERCIAL READINESS

CONTACT

National Institute of Advanced Industrial Science and Technology (AIST)
1-1-1 Umezono, Tsukuba,
Ibaraki 305-8560, Japan
+81 298612000
www.aist.go.jp

RESPONSIVE MEMBRANE WINDOW

Vascular Skin

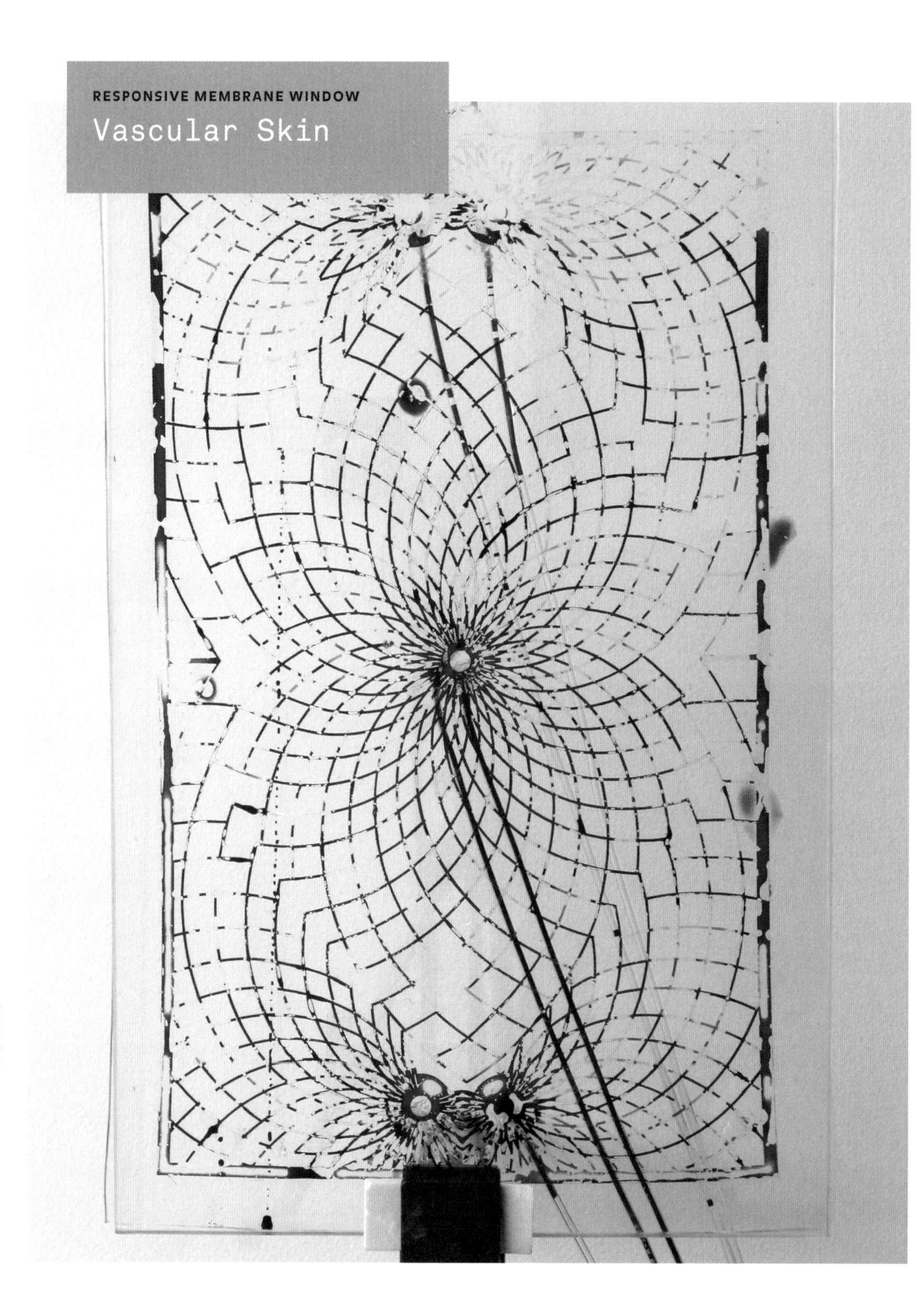

Hydrogels are polymeric materials capable of retaining large quantities of water within three-dimensional networks. Researchers at the MIT Media Lab sought to explore the hydrophilic-hydrophobic phase transition in hydrogels, which occurs slightly above room temperature, in a new form of glazing. The Vascular Skin window demonstrates the mechanical and optical transformations that happen at this phase transition, including permeability, swelling, and visual modification—thus allowing light and view control that responds to the ambient temperature. In this way, the Vascular Skin project presages the possibility of fully active building surfaces and products.

CONTENTS
Glass, hydrogels, tubing

APPLICATIONS
Windows, glazed partitions, temperature-responsive surfaces

TYPES / SIZES
Vary

ENVIRONMENTAL
Opportunities for passive solar thermal control

FUTURE IMPACT
Building envelopes that imitate the capillary behavior of biological skin

COMMERCIAL READINESS

CONTACT
Mediated Matter Group
Massachusetts Institute of Technology Media Lab
75 Amherst Street, Room E14–433B, Cambridge, MA 02139-4307
617-324-3626
matter.media.mit.edu
ked03@media.mit.edu

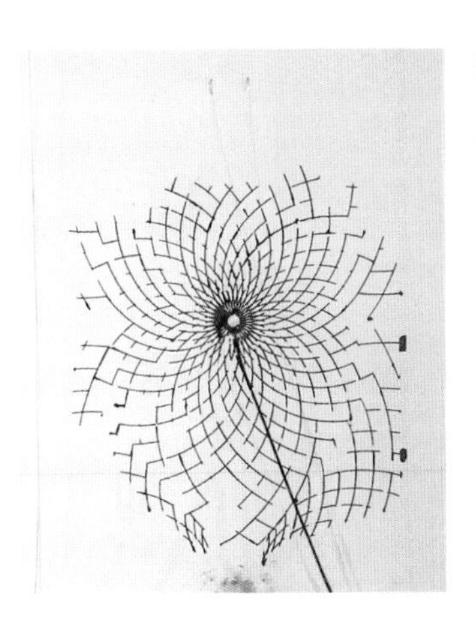

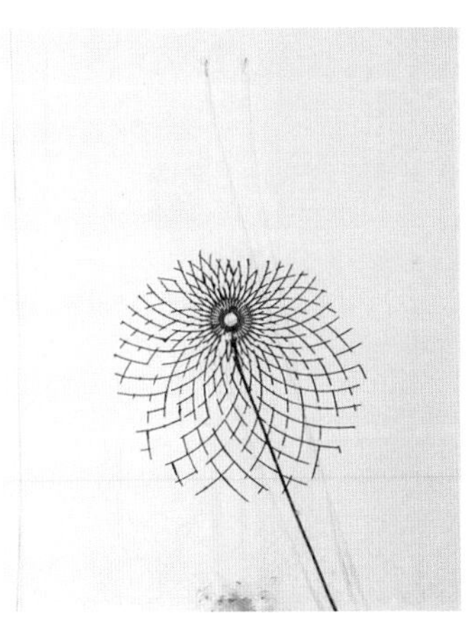

7

Paint and Coatings

> Living Surfaces

> *Trompe l'oeil*, meaning "deceive
> the eye," is a technique for
> painting the illusion of a three-
> dimensional object or environment
> on a two-dimensional surface.
> The term emerged during the
> seventeenth century from baroque
> artists' fascination with illusory
> optical effects; however, the
> practice is known to date back at
> least as far as ancient Greece,
> when it was commonly used in
> murals. Trompe l'oeil represents
> a symbolic bridge between archi-
> tecture and art, insofar as it
> connects reality with fantasy, or
> the corporeal world with the realm
> of the imagination.

In trompe l'oeil, we see the two familiar, distinct modes of painting fused: coating a surface, as when painting a wall; and constructing a work of art, as in creating a mural. René Magritte famously exposed the inherent difference between these approaches in his 1929 painting *The Treachery of Images*. A realistically rendered pipe is depicted with the caption *Ceci n'est pas une pipe*, or "This is not a pipe." The first-time viewer's confusion is soon replaced by comprehension: this is not an actual pipe, but merely its representation.

Paint is the vehicle that makes such an illusion possible.[1] Its application has a deep history—in fact, paint is the oldest known substance of design. The material of Paleolithic cave art made in Asia and Europe some forty thousand years ago, and discovered in a one-hundred-thousand-year-old paint-making studio, paint was used long before any known form of building.[2] For millennia, paints, dyes, and other material coatings were employed largely as visual tools, their ingredients mixed by hand. In the early eighteenth century, pigment-grinding machines were developed to facilitate the process of achieving particular colors.[3] Industrialization expanded the availability of relatively inexpensive coatings, and paint became more commonly used in building applications, such as interior house painting. Today paint has various functions in the physical environment, including coating vulnerable surfaces and creating uniform fields of color. Yet long before paint was used as a protective and homogenizing treatment, it served to record and communicate ideas.

Next-generation paints revive the emphasis on communicating ideas, by way of enhanced material capacities. Emerging coating technologies include exotic substances like quantum dots, graphene, and carbon-sequestering dyes, and enable functions like light harvesting, electricity conduction, and structural monitoring. Considered in combination with new automation processes, these technologies demonstrate that paint is no longer an inert substance, but a fluid channel for sensing and response. Paint's roles are now rapidly expanding, as more functions are packed into an ultrathin veneer.

Researchers at the University of Strathclyde in Glasgow, for example, have developed a smart paint that detects and communicates stresses in structures before they become problematic (see page 216). Lead scientist Mohamed Saafi pursued the idea based on the need for constant maintenance in wind turbines. According to Saafi, "Wind turbine foundations are currently being monitored through visual inspections. The developed paint with the wireless monitoring system would significantly reduce the maintenance costs and improve the safety of these large structures."[4] Composed of fly ash and aligned carbon nanotubes, the conductive coating is simply spray applied and connected to electrodes for full operation.

← To enhance functionality, carbon nanotubes have been incorporated into a variety of coatings, such as inks that may be printed to create nanosensors (see page 212). Other smart paints include printable batteries made of solid-state electrolyte paste, a solar-harvesting paint made from semiconductor nanocrystals, and a color-shifting crystalline fluid that enables active camouflage.[5]

Surface treatments have become important agents in environmental remediation. For industrial processing applications, scientists have discovered that CO_2 may be employed in lieu of a toxic solvent to impregnate polymers with pigment (see page 202). ➔ **Italian manufacturer Italcementi has developed a cement with photocatalytic ability.[6] With the addition of titanium dioxide (TiO_2), the cement is able to break down local air pollution in the presence of direct sunlight.[7]** According to one company study, TiO_2-impregnated concrete blocks reduced 45 percent more NO_x (nitrogen oxide caused by combustion) than conventional asphalt paving.[8] Designers have also employed pollution-responsive dyes in the service of making ecological provocations. For example, London-based artist Lauren Bowker created the $PdCl_2$ jacket by treating fabric with palladium chloride ink, which is typically used in carbon monoxide detection (see page 210). In the presence of CO_2-heavy air, the jacket transforms from yellow to black, thus making visible an otherwise invisible environmental condition.

Automation is also changing the field of paints and coatings. The world's first commercial painting robot was made in Trallfa, Norway, in 1969.[9] Used frequently in industries such as automotive and aerospace manufacturing, robotic paint systems are advantageous in that they can handle a variety of paint processes and are resistant to toxic—and sometimes even explosive—environments filled with solvents.[10] ➚ **The Germany-based company Sonice Development creates robotic paint systems for architectural applications. Its mobile robots are not confined to factory environments, but are designed for use on-site. The firm's colorspace robots are machines that climb vertical surfaces while applying multicolored paint according to predetermined algorithms (see page 204).** Its Facadeprinter is just that—a software-controlled robot that creates pointillist images on expansive building envelopes (see page 208).

New paints, dyes, inks, and various coating methods represent unprecedented capabilities in surfacing applications. As a result, they change our relationship with

surfaces in general. Until recently, paint has been considered an inert substance. Yet it is becoming increasingly dynamic and responsive. For instance, ← **materials like prismatic pigments and thermochromic dyes—which exhibit transformative effects depending on a viewer's shifting position or changing temperature, respectively—have by now become familiar.** The next wave of advanced surfacing materials and processes offers additional capacities for sensing and responding, energy harvesting and storage, environmental remediation, and automation. Thus paint, the age-old material bridge between subject and object, is now expanding its operations within the designed environment—and enriching the possible connections between reality and representation.

---

1 I use *paint* here in general reference to coatings, pigments, dyes, inks, and other colorants and surface treatments.

2 A. W. G. Pike et al., "U-Series Dating of Paleolithic Art in 11 Caves in Spain," *Science* 336, no. 6087 (June 15, 2012): 1409, http://science.sciencemag.org/content/336/6087/1409. See also accessed January 18, 2016, http://www.livescience.com/16538-oldest-human-paint-studio.html. By contrast, the earliest known house is only about 11,500 years old. See accessed January 18, 2016, http://www.telegraph.co.uk/news/uknews/7937240/Oldest-house-in-Britain-discovered-to-be-11500-years-old.html.

3 James Ayres, *Art, Artisans and Apprentices: Apprentice Painters & Sculptors in the Early Modern British Tradition* (Oxford: Oxbow Books, 2014), 117.

4 "Smart paint could revolutionize structural safety," University of Strathclyde Glasgow, accessed January 18, 2016, https://www.strath.ac.uk/research/subjects/civilenvironmentalengineering/smartpaintcouldrevolutionisestructuralsafety/.

5 See accessed January 18, 2016, http://pubs.acs.org/doi/abs/10.1021/acs.nanolett.5b01394; accessed January 18, 2016, http://phys.org/news/2011-12-cheap-quantum-dot-solar.html; and accessed January 19, 2016, http://ns.umich.edu/new/multimedia/videos/22148-chameleon-crystals-could-make-active-camouflage-possible.

6 See accessed January 18, 2016, http://www.italcementigroup.com/NR/rdonlyres/1F30E487-C0A2-4D6F-AB6D-C14555FD866F/0/Scientificresults.pdf.

7 Concerns have increased over the potential toxicity of titanium dioxide, which is largely a factor of size. Particles larger than the nano scale (100 nm in diameter or less), such as conventional pigment-grade powders, are generally recognized as safe. See accessed January 18, 2016, http://phys.org/news/2015-03-dunkin-donuts-ditches-titanium-dioxide.html.

8 Italcementi Group, "TX Active: the active photocatalytic principle" (manufacturer report), accessed January 18, 2016, 6.

9 Mike Wilson, *Implementation of Robot Systems: An introduction to robotics, automation, and successful systems integration in manufacturing* (Oxford: Butterworth-Heinemann, 2014), 8.

10 Ibid., 56.

COLOR-CHANGING MICROALGAE DYE

Algaemy

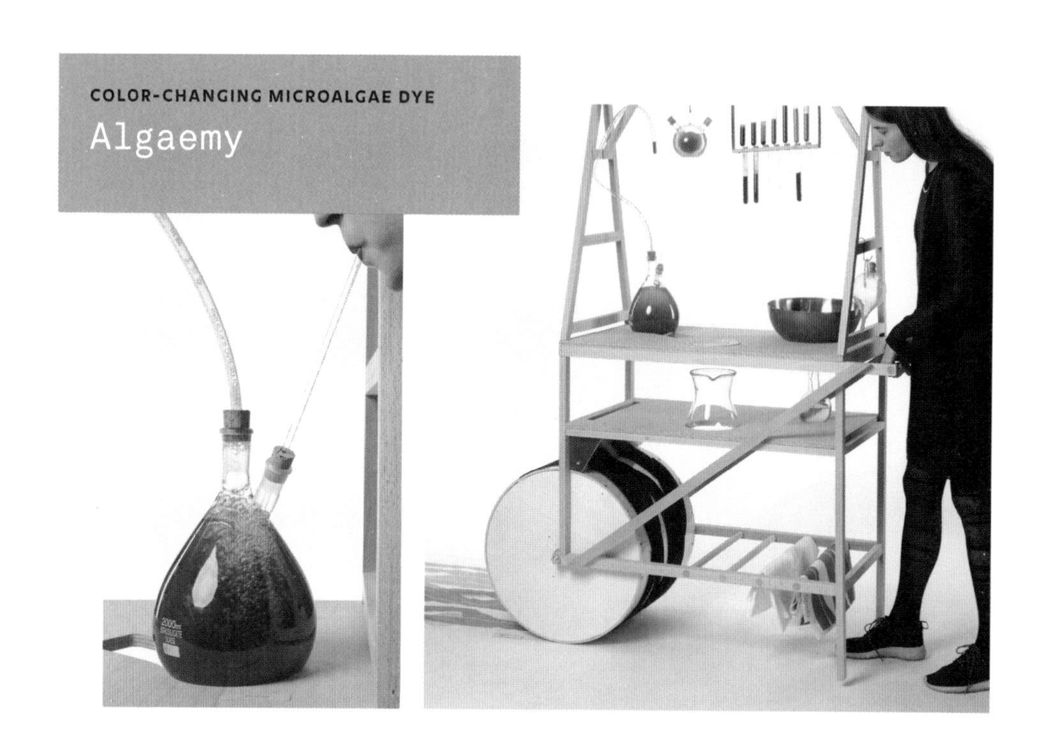

Algaemy is a project that investigates the potential of microalgae as a pigment in textile printing. All patterns consist of living, self-harvested material. The result of a research collaboration between Blond & Bieber and the Fraunhofer Institute for Interfacial Engineeing and Biotechnology (IGB), the project consists of a microalgae-based dye palette developed to create environmentally responsible textiles.

Because the colors are not lightfast, they transform over time, unlike conventional, chemical-based textile dyes. Algaemy's biodynamic colors gradually change when exposed to sunlight. For example, green becomes an intense blue, while pale pink turns bright red and eventually orange. Every textile thus conveys a temporal story based on exposure and use. In this way, Algaemy illustrates the potential of an autonomously self-grown pigment created from an underappreciated natural substance.

CONTENTS
Microalgae, printing paste

APPLICATIONS
Cotton, leather, paper, and other surfaces

TYPES / SIZES
Natural color palette based on different microalgae species

ENVIRONMENTAL
100 percent natural and biocompatible pigment, nontoxic and biodegradable substance

LIMITATIONS
Currently limited to small-batch production

FUTURE IMPACT
Potential to replace many chemical-based pigments conventionally used in the textile industry with a 100 percent natural and benign alternative

COMMERCIAL READINESS

CONTACT
Studio Blond & Bieber
Gounodstrasse 26, Berlin 13088, Germany
+49 15773115527
www.blondandbieber.com
info@blondandbieber.com

CO_2 INFUSION-BASED PLASTIC COATING

CO_2-Impregnated Polymers

PROCESS

CO_2 has a negative reputation for its significant contribution to global warming. However, CO_2 also exhibits some positive features that may be leveraged in product manufacturing. For example, the gas may substitute for environmentally problematic solvents conventionally used in paints and coatings. Nontoxic, nonflammable, and widely available, CO_2 can be utilized to impregnate plastics with durable, scratch-resistant coatings that are superior to paint.

Scientists at the Oberhausen, Germany–based Fraunhofer Institute for Environmental, Safety, and Energy Technology (UMSICHT) have developed a process to embed CO_2 within polymers, such as polycarbonate and nylon. First, they heat the gas to 30.1 °C at a pressure of 73.8 bar, imparting it with solvent-like characteristics. In this state, it functions as a vehicle for dyes and other additives. Increasing the pressure to 170 bar allows powdered pigment to be dissolved completely into the CO_2 and subsequently into its plastic carrier. The process requires only a few minutes and completely encapsulates the pigment within the plastic.

In addition to being used with durable coatings, the method may also be used with additives, such as antibacterial coatings and medical compounds. The researchers conducted tests demonstrating that CO_2-impregnated polycarbonate is suitable for mobile device casings or door handles that eliminate *E. coli* bacteria, for example. They also speculate that the process could be used to manufacture colored contact lenses that release pharmaceutical compounds over a controlled period.

CONTENTS

Polycarbonate or another noncrystalline polymer, liquid CO_2, pigment

APPLICATIONS

CO_2-based polymer pigmentation; manufacture of plastics with scratch-resistant, fire-retardant, or UV-stable coatings

TYPES / SIZES

May be used with polycarbonate, nylon, TPE, TPU, PP, or other amorphous or partially crystalline plastics

ENVIRONMENTAL

Puts atmospheric CO_2 to a functional use, ecologically responsible alternative to solvents used in paints and other coatings

TESTS / EXAMINATIONS

Laboratory tests

LIMITATIONS

Cannot be used with highly crystalline polymers

FUTURE IMPACT

Utilization of CO_2 as an environmentally friendly coating material

COMMERCIAL READINESS

● ○ ○ ○ ○

CONTACT

Fraunhofer Institute for Environmental, Safety, and Energy Technology UMSICHT
Osterfelder Strasse 3, Oberhausen 46047, Germany
+49 208-85980
www.umsicht.fraunhofer.de
info@umsicht.fraunhofer.de

ROBOTICALLY CRAFTED MURALS

Colorspaces

PROCESS

Colorspaces are robotically crafted drawing installations devised and fabricated by Berlin-based Sonice Development. An autonomous machine with the capacity to move on vertical surfaces applies paint to a designated area using a marker. Over a time span of hundreds of hours, the application of different colored lines forms a dense, multispectral field. The meandering behavior of the robot corresponds to simple algorithmic rules with a few random elements, based on custom-developed software. As a result, the multicolored mural continuously changes as long as the robot remains active.

Sonice Development has created four types of colorspaces for different site-specific installations: Emerging Colorspace, Sensing Colorspace, Rising Colorspace, and Crosshatching Colorspace.

CONTENTS

Surface-mounted robotic system, custom hardware and software, water-based acrylic paint

APPLICATIONS

Commissioned artworks, surface treatments

TYPES / SIZES

Emerging Colorspace, Sensing Colorspace, Rising Colorspace, and Crosshatching Colorspace; minimum wall dimensions 13' 1.5" × 13' 1.5" (4 × 4 m)

LIMITATIONS

For interior applications

FUTURE IMPACT

Robot-driven paint and finish applications on vertically oriented surfaces, algorithm-based on-site pattern generation

COMMERCIAL READINESS

CONTACT

**Sonice Development GmbH
Glogauer Strasse 21, Berlin D-10999, Germany
www.sonicedevelopment.com
contact@sonicedevelopment.com**

PROGRAMMABLE COLOR INK TILES

E Ink Prism

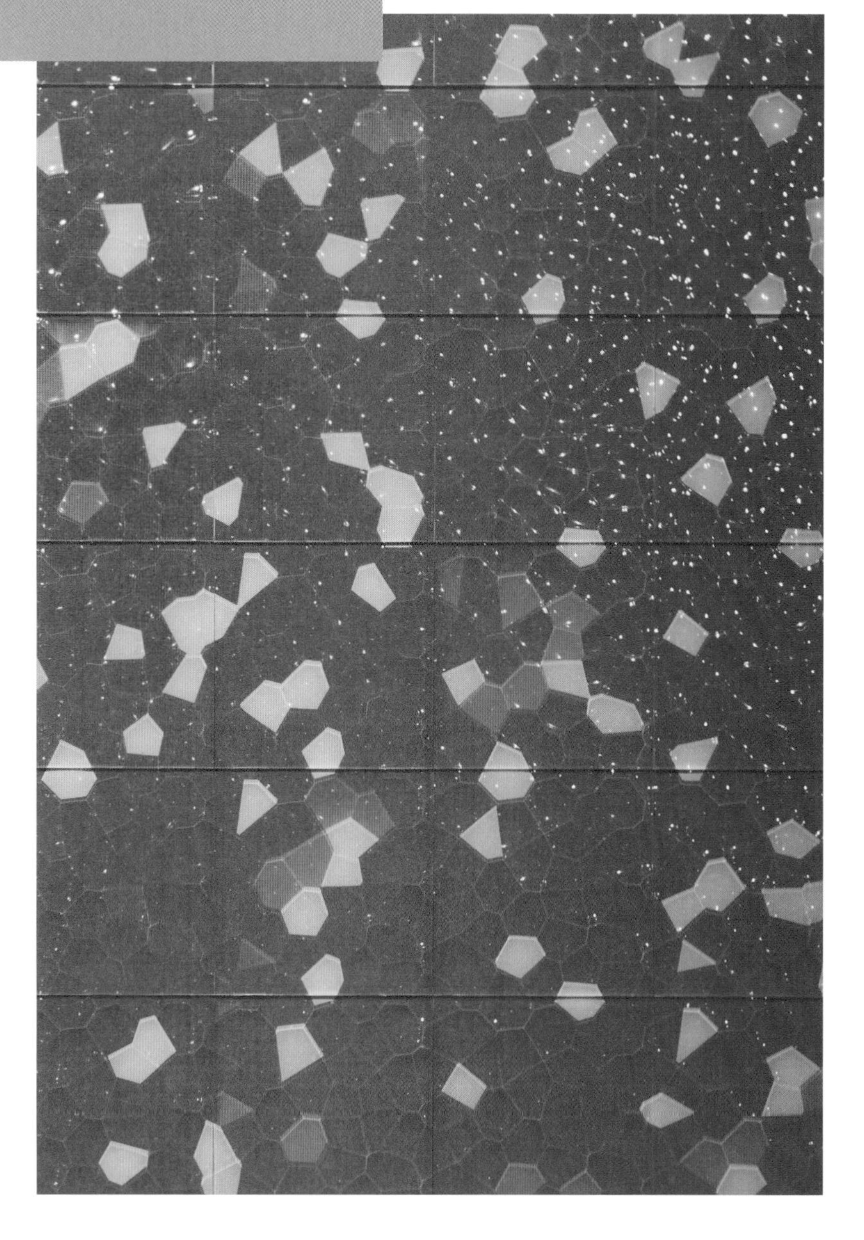

PRODUCT

Buildings consist primarily of materials intended to be static and homogeneous. Meanwhile, digital interfaces are typically used episodically in interior spaces, as focal points for interactivity. However, recent developments in lighting and digital technologies enable the fusion of digital interfaces and immersive environments, permitting the creation of continuously transforming architectural surfaces.

E Ink Prism is a tile-based technology that utilizes bistable electronic ink film. The film consists of printing industry pigments that are similar to paint, yet the pigment can be made to appear or disappear as desired—enabling an immediate color transformation of an entire wall or room with the flip of a switch. The durable, low-power tile system expands the capabilities of simple digital interfaces beyond conventional screen-based devices, transforming spaces into graphic or textual information canvases.

CONTENTS

Electronic ink film, colored pigments, multilayered plastic tile

APPLICATIONS

Interactive walls, programmable surfaces, signage, displays, bichromatic screens

TYPES / SIZES

Bichromatic tile, standard size 4 × 4 × $^{1}/_{2}$" (10 × 10 × 1.3 cm)

ENVIRONMENTAL

Low power consumption

LIMITATIONS

Limited to two colors in graduated hues

FUTURE IMPACT

Responsive displays without conventionally high energy consumption, dynamic and programmable environments

COMMERCIAL READINESS

CONTACT

E Ink Holdings Inc.
3 Li-Hsin Road 1, Hsinchu Science Park,
Hsinchu 300, Taiwan
+886 35798599
www.eink.com
info@eink.com

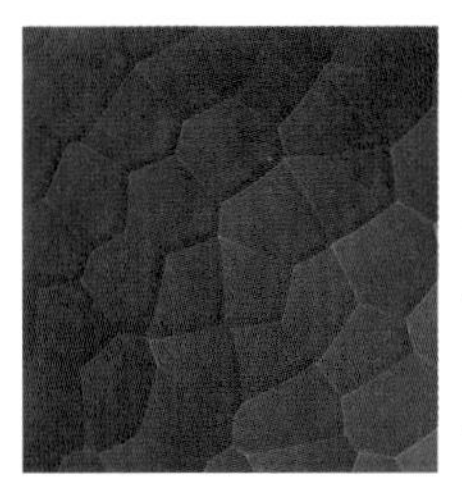

ARCHITECTURAL-SCALE ROBOTIC PRINTER

Facadeprinter

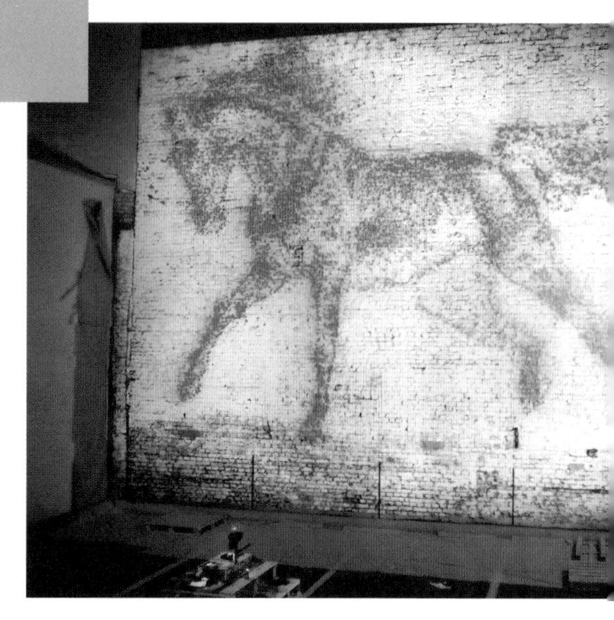

PROCESS

The Facadeprinter creates large murals by firing paintballs onto walls. The software-controlled robot consists of a two-axis turntable and a pressurized air-based printhead. The printer "shoots" large-scale images onto a surface at a predetermined distance, based on programmed patterns. According to the manufacturer, the Facadeprinter is the world's first distance printer. A tool for generating large-scale media, the device is akin to a dot matrix printer of architectural dimensions.

CONTENTS

Robotic paint projector, custom hardware and software

APPLICATIONS

Mural painting, signage, public performances

TYPES / SIZES

Printer position and dot density may be adjusted on site; minimum print size 13 × 13' (4 × 4 m), maximum print size 26 × 33' (8 × 10 m)

TESTS / EXAMINATIONS

Site-based installations since 2009

LIMITATIONS

Paint used is not UV stable, printing distance is 13–23' (4–7 m)

FUTURE IMPACT

Increased use of robotics in on-site building applications and performance art

COMMERCIAL READINESS

CONTACT

Sonice Development GmbH
Glogauer Strasse 21, Berlin D-10999, Germany
www.sonicedevelopment.com
contact@sonicedevelopment.com

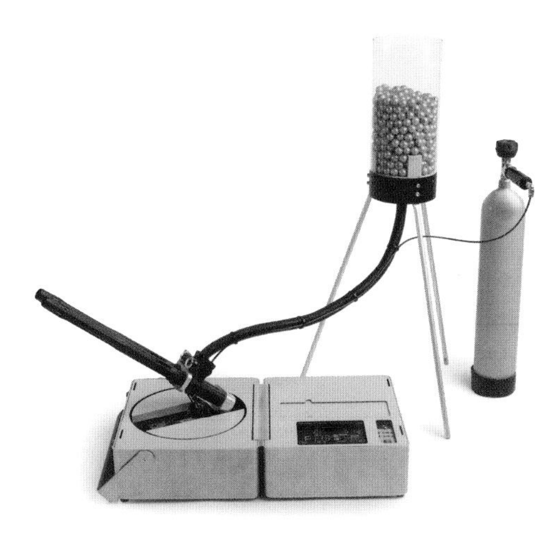

099113-002

POLLUTION-TRIGGERED DYE

$PdCl_2$

MATERIAL

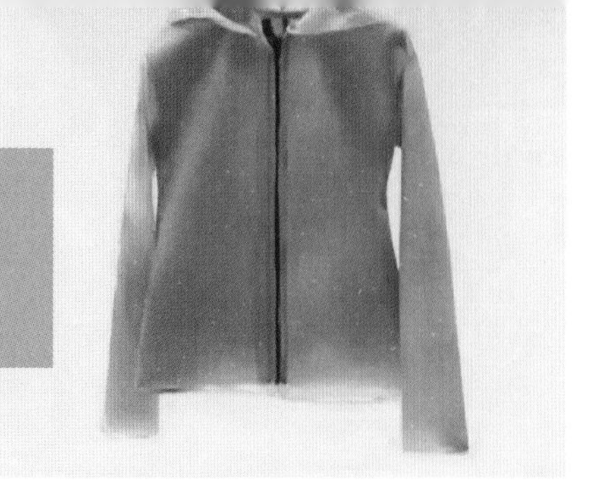

Palladium chloride, or $PdCl_2$, is a palladium-based compound used in carbon monoxide detectors and to test corrosion resistance in stainless steel. The chromic ink responds to the presence of CO_2 emissions in the atmosphere, transforming when the CO_2 rises above a certain level in a reversible color change from yellow to black. London-based fashion designer Lauren Bowker developed a dye from $PdCl_2$ for fabric applications and created a pollution-absorbing jacket that communicates the presence of carbon emissions.

Bowker's invention speaks to an increasing awareness of global environmental hazards, including not only greenhouse gas emissions but also human health–related toxins, like secondhand smoke. According to the designer, the $PdCl_2$ is a visual platform that explores the aesthetic dimensions of environmental health concerns. Her invention uses a language of visual communication to reveal invisible hazards of our surroundings.

CONTENTS

$PdCl_2$ fabric dye

APPLICATIONS

Clothing, awnings, canopies, and other textile fabrications

TYPES / SIZES

Reversible bicolor transition between yellow and black

ENVIRONMENTAL

Alerts viewers to the presence of rising CO_2

FUTURE IMPACT

Wearable textiles and fabric architecture that function as litmus tests for various environmental hazards

COMMERCIAL READINESS

CONTACT

The Unseen
Waterside 44-48 Wharf Road,
London N1 7UX, UK
+44 (0) 2078363165
www.seetheunseen.co.uk
contact@seetheunseen.co.uk

PRINTABLE SENSORS THAT INDICATE MATERIAL DEFORMATION

Printable Nanosensors

Yeadon Space Agency creates electrically conductive composite materials using carbon nanotube–based inks. The nanotube coatings are durable and moisture resistant, and they retain their conductive capacity when stretched or strained, unlike many metal coatings. Yeadon Space Agency introduces nano ink to films, textiles, concretes, and other materials to make composites that function as deformation-monitoring sensors. When electrical current runs through the nanotubes, any material strain changes their resistance, and this change signals that the composite is experiencing deformation. With their responsive capacity, Printable Nanosensors offer the potential to create flexible, lightweight, electrically conductive products that respond to environmental cues via ubiquitous sensing.

CONTENTS

Single-wall carbon nanotubes, 1.5 nm, 90 wt percent purity; deionized water; sodium dodecyl sulfate; substrate (textile, concrete, etc.)

APPLICATIONS

Apparel and other manufactured products, industrial design, architecture, interior design

TYPES / SIZES

Vary

FUTURE IMPACT

Advancements in interactive products and environments will increasingly require lightweight, flexible, electrical networks with remote, ubiquitous sensing

COMMERCIAL READINESS

● ○ ○ ○ ○

CONTACT

Yeadon Space Agency
33 Flatbush Avenue, 6th Floor, Brooklyn, NY 11217
www.yeadonspaceagency.com
agent@yeadonspaceagency.com

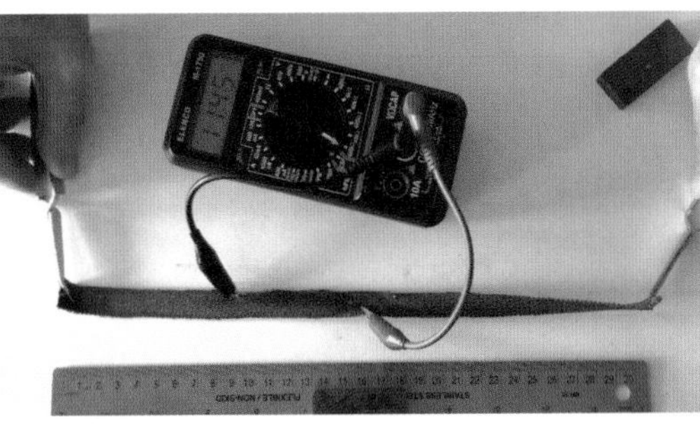

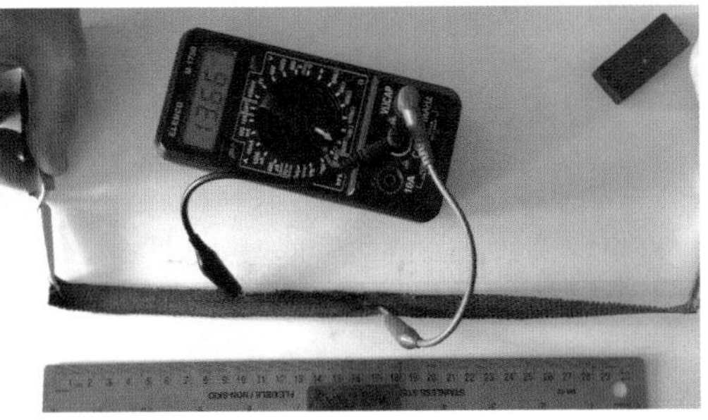

BIORESIN-COATED RECEPTACLES

Propolis Vessels

Propolis is a biodegradable resin collected by honeybees for patching beehive openings. Its color varies with its particular botanical origins, although brown is the most typical hue. French designer Marlène Huissoud works with a black propolis from rubber trees, which is a mix of between 50 and 150 different waxes, balsams, resins, pollens, and essential oils.

Because black propolis is similar to glass, Huissoud created a collection of vessels using a variety of glass techniques. After many experiments, she succeeded in blowing the material using standard glassblowing methods. She had to adapt the kiln for the propolis, however, since its melting point of 212 °F (100 °C) is much lower than that of glass, which is about 2,550 °F (1,400 °C). Inspired by the material's arboreal origin, the completed vessels represent tree trunks, and Huissoud engraves each one with a different pattern.

CONTENTS

95 percent honeybee bioresin, 5 percent oak

APPLICATIONS

Sculptural objects and vessels, interior surfaces

TYPES / SIZES

Viscosity varies with different ingredients; vessels 4–10" (10–25 cm) diameter, 10–39" (25–100 cm) tall

ENVIRONMENTAL

Biocompatible surfacing material, biodegradable

LIMITATIONS

Low melting point, for interior use only

FUTURE IMPACT

Promising natural alternative to toxic industrial sealants and coatings

COMMERCIAL READINESS

CONTACT

Studio Marlène Huissoud
34C Mattock Lane, London W55BH, UK
+44 (0) 7445310774
www.marlene-huissoud.com
info@marlene-huissoud.com

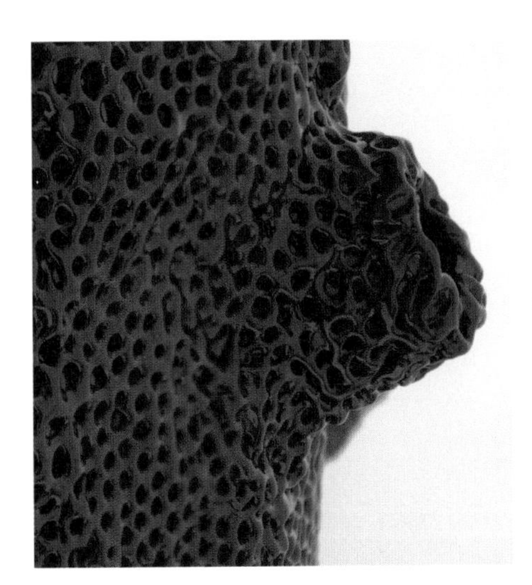

PAINT THAT MONITORS STRUCTURAL HEALTH

Smart Paint

Scientists at Glasgow's University of Strathclyde have developed a Smart Paint for monitoring the physical health of the structures it coats. The paint consists of a mixture of fly ash, a by-product of coal combustion, and highly aligned carbon nanotubes. The combination results in a cement-like coating that can be used to monitor structural stresses and the development of microscopic faults, as the carbon nanotubes become more or less conductive in the presence of corrosion and stress, respectively. The paint not only makes use of an industrial waste material, but also functions well in harsh environments. It is therefore well suited for use in coating structures such as bridges, tunnels, mines, and wind turbines, which must be designed to withstand heavy use and weathering.

Traditional approaches to assessing the physical condition of large structures are expensive and time-consuming. The researchers estimate that Smart Paint costs a mere 1 percent of traditional inspection methods, resulting in significant long-term savings. The material can also communicate early warning signs of physical problems before they become apparent in the form of structural damage.

CONTENTS

Recycled fly ash, carbon nanotubes, binding agents, wireless communication nodes

APPLICATIONS

Bridges, tunnels, mines, wind turbines, and other infrastructure works; building structures and envelopes

TYPES / SIZES

Vary

ENVIRONMENTAL

Reuse of an industrial by-product of coal combustion, optimizes material longevity

LIMITATIONS

Subject to availability of fly ash

FUTURE IMPACT

Ability to diagnose physical health of large structures inexpensively

COMMERCIAL READINESS

● ○ ○ ○ ○

CONTACT

University of Strathclyde
Civil & Environmental Engineering
75 Montrose Street, Glasgow G1 1XJ, UK
+44 (0) 1415483275
www.strath.ac.uk
contact-civeng@strath.ac.uk

LIQUID FILM WITH SWITCHABLE PROPERTIES

Tunable Material

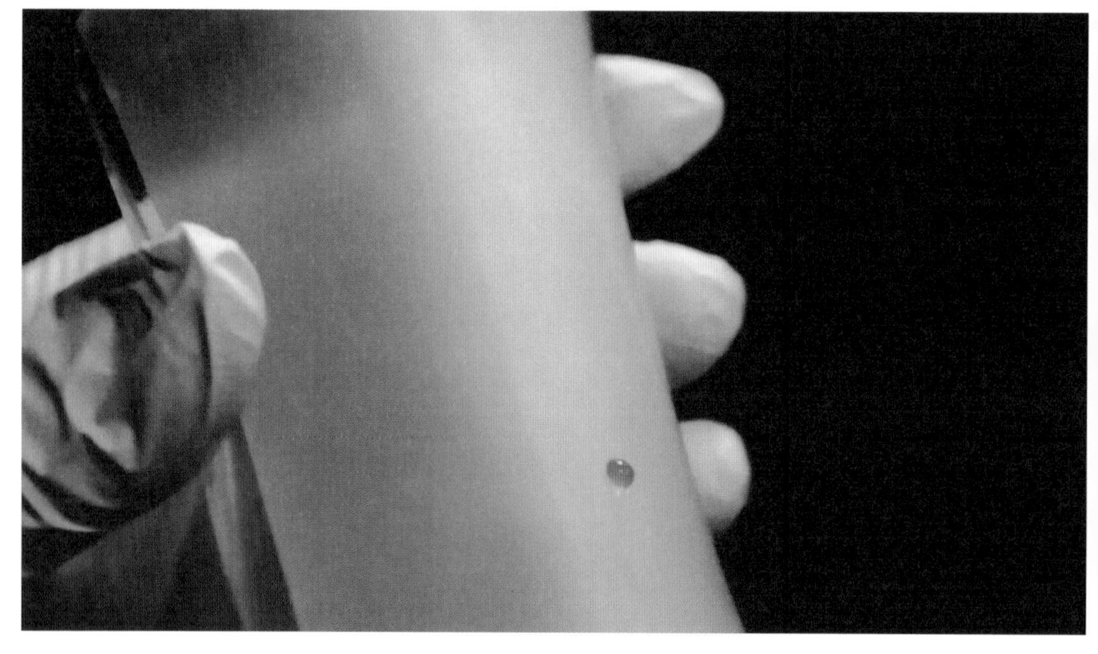

Dynamic natural processes are inspiring a new generation of adaptive material technologies. The tears that coat human eyes, for example, are fascinating for their ability to maintain optical clarity, provide protection from foreign particles, and flush out wastes. Taking inspiration from tears, scientists at Harvard's Wyss Institute for Biologically Inspired Engineering and the university's John A. Paulson School of Engineering and Applied Sciences have developed a continuous liquid coating with highly controllable capabilities.

The new material consists of an elastic substrate embedded with micropores, coated with a liquid film. When the substrate is stretched or bent, the pore sizes change, causing deformations in the liquid coating. Variations in the film exhibit precise controllability over two characteristics: transparency and wettability. When the material is at rest, it is smooth, light transmitting, and hydrophobic. In tension, it becomes rough, opaque, and hydrophilic. The coating may also be used to adjust adhesion or antifouling characteristics.

Future applications of the highly tunable film include responsive building envelopes, plumbing that regulates liquid flow rate automatically, and fabrics that can adjust their opacity and hydrophobicity based on weather conditions.

CONTENTS
Elastic porous substrate, continuous liquid film

APPLICATIONS
Building facade coatings, oil and gas pipelines, responsive textiles, microfluidic and optical systems

TYPES / SIZES
Vary

ENVIRONMENTAL
Controllable self-cleaning and protective capabilities

TESTS / EXAMINATIONS
Laboratory tests

FUTURE IMPACT
Responsive surfaces that enable precise control over opposing physical properties

COMMERCIAL READINESS
● ○ ○ ○ ○

CONTACT
Wyss Institute for Biologically Inspired Engineering
Harvard University
3 Blackfan Circle, Boston, MA 02115
617-432-7732
www.wyss.harvard.edu
info@wyss.harvard.edu

PHOTOLUMINESCENT PAVING SURFACE

Van Gogh Path

PRODUCT

The Smart Highway project, developed by Studio Roosegaarde and Heijmans, envisions responsive roadways that employ energy, light, and data to create safer and more informative driving experiences. The team built a test roadbed with photo-luminescent stripes that absorb daylight and emit light during the evening for up to eight hours.

The project also includes a version tailored to pedestrians and bicyclists. The Van Gogh Path is composed of thousands of fluorescent stones, which are embedded in paving in spiraling patterns to invoke the artist's celebrated painting *The Starry Night*. Like the Glowing Lines roadbed, the Van Gogh Path utilizes sunlight for passive nocturnal illumination. One of the first paths was installed in Neunen, Netherlands, the place where Van Gogh lived in the late nineteenth century. With swirling constellations that mirror the night sky, the path is an intelligent marriage of public art and infrastructure.

CONTENTS

Photoluminescent paint, aggregate

APPLICATIONS

Bicycle paths, sidewalks, and other circulatory infrastructure

TYPES / SIZES

Green or blue fluorescent aggregate

ENVIRONMENTAL

Zero-energy nocturnal illumination

TESTS / EXAMINATIONS

Pilot installations at Neunen, Eindhoven, and Maastricht, Netherlands

LIMITATIONS

Low-level illumination is not as powerful as conventional street lighting

FUTURE IMPACT

Net-zero illumination of pathways and roadways, fusion of public infrastructure and art

COMMERCIAL READINESS

CONTACT

Studio Roosegaarde
Vierhavensstraat 52–54,
Rotterdam 3029 BG, Netherlands
+31 103070909
www.studioroosegaarde.net
mail@studioroosegaarde.net

8 Fabric

> Generative Skin

> Textiles are often compared to
> skin, in the sense that they
> function as a second skin to the
> body in the form of clothing,
> and have at times originated from
> animal skins. Architecture can
> be considered a skin as well.
> As textile designer Anni Albers
> wrote in "The Pliable Plane,"
> "If we think of clothing as a
> secondary skin we might enlarge
> on this thought and realize that
> the enclosure of walls in a
> way is a third covering, that our
> habitation is another 'habit.'"[1]
> Albers's analogy does more than
> simply articulate the relative
> physical position and importance
> of clothing and shelter; it
> serves to bridge the two.

At the time of the article's publication in 1957, she argued that the role of textiles in architecture was underappreciated. "The essentially structural principles that relate the work of building and weaving could form the basis of a new understanding between the architect and the inventive weaver," she asserted. "Textiles, so often no more than an after-thought in planning, might take place again as a contributing thought." [2]

Over half a century later, textiles have fulfilled this potential. Advanced fabrics and fiber composites are now widely employed in the forms of elaborate canopies, mast-supported structures, shade systems, and textile-based facades. Propelled by fabric architecture pioneers like Frei Otto, who founded the influential Institute for Lightweight Structures at the University of Stuttgart in 1964 and designed the 1972 Munich Olympic Stadium and many other significant projects, textile materials and structures have made calculable advances in the ensuing decades. Meanwhile, textiles' connective capacity continues to expand. Just as fabric technologies have traversed the second and third skins, so they are now joining the first two. Biomimetics and biodesign have inspired new explorations of living structures and processes, with burgeoning efforts to align multiscalar textile design with the mechanics of natural systems. According to architecture professor Mark Garcia, "Complex functions and qualities like self-repair, growth and replication are the organic qualities of architecture being sought after within this model of research currently shared by architects, textile designers, engineers and other scientists." [3] Emerging fabric technologies reveal an intensified interest in tissue-like structured surfaces, protein-emulating fibers, tactile interactivity, and biomechanical response.

London-based designer Julian Melchiorri's experiments illustrate this focus. In 2014 Melchiorri developed a biotextile consisting of silk fabric impregnated with plant cell–derived chloroplasts. The so-called Silk Leaf fuses silk mesh and living, photosynthesizing cells. "As an outcome I have the first photosynthetic material that is working and breathing as a leaf does," declared the designer.[4] Like Melchiorri, **↓ New York–based The Living employs synthetic biology to broaden the design palette. The firm's Bio-Fabrication process generates multiproperty composite materials from different strands of bacteria** **(see page 230).** The development

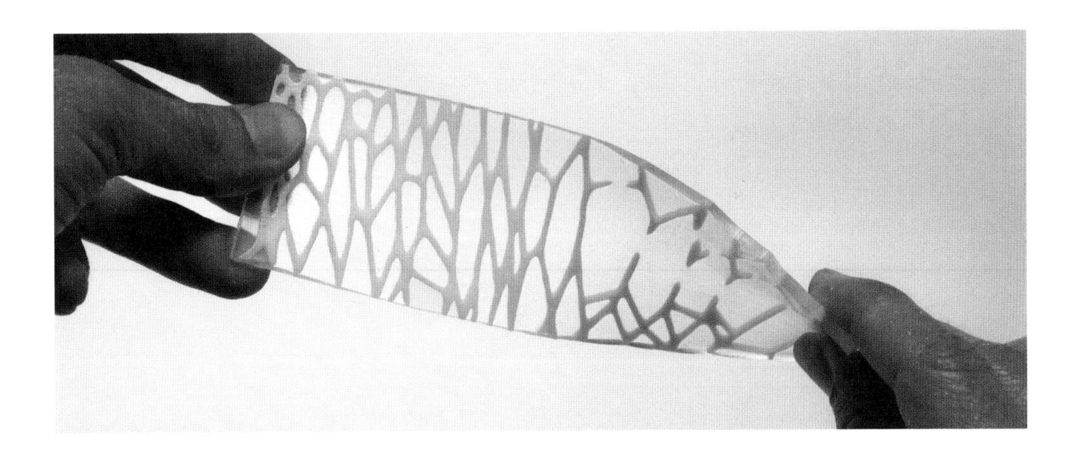

of superbiotextiles includes ultraperforming fibers like Qmonos, made to emulate spider dragline silk based on silk's superior strength and environmental performance over steel (see page 246). Japanese manufacturer Spiber produces the fiber using bioengineered bacteria and recombinant DNA and is launching the first article of clothing made of synthetic spider silk. Other biotextile research highlights the "first skin," such as a new liquid skin composite created to restore damaged tissue or an artificial membrane that responds to the presence of harmful solvent vapor (see page 248).

Research focused on bioresponsive effects has led to a variety of promising new textile technologies. ↗ **Australia-based Commonwealth Scientific and Industrial Research Organisation's smart bandage enhances healing and wound management by employing thermochromic fibers that indicate infection and other healing problems (see page 250).** Electronic tattoos likewise enable sophisticated biomonitoring, and the ultrathin, flexible microelectronics are much less cumbersome than conventional electrocardiograms and other bulky medical devices.[5] Researchers at the Korea Advanced Institute of Science and Technology have developed a wearable battery that is flexible and rechargeable, overcoming the hurdles presented by the delicate wiring and rigid, bulky components of most wearable electronics.[6] Building on this trend, materials scientist Yoel Fink and his team at MIT's Research Laboratory of Electronics are developing active textiles based on their belief that fabrics can do more. In 2010 they created fibers that can detect and produce sound—and by extension textiles that act as microphones and speakers, with threads that generate power when stretched.[7] EJTech's Chromosonic is a similarly interactive fabric, and its programmable surface changes color in response to changes in sound (see page 234).

Biology has also inspired the creation of dynamic structured textiles that adapt to shifting conditions. For example,

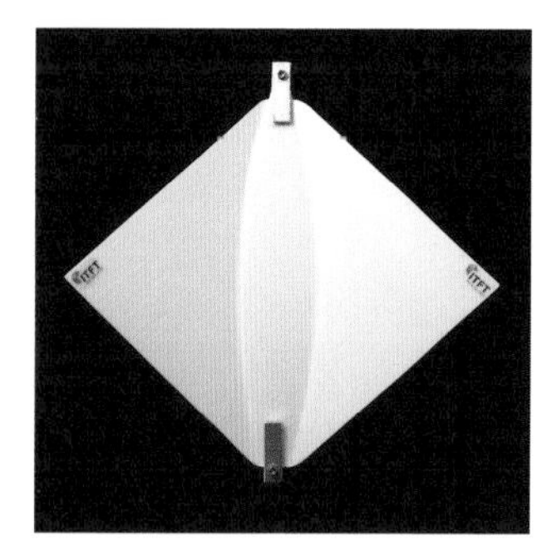

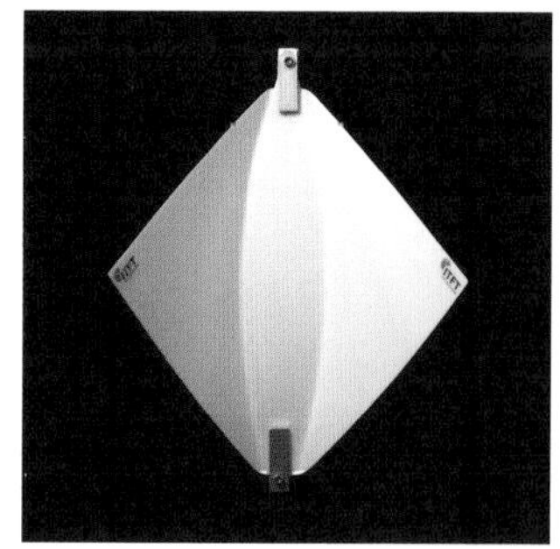

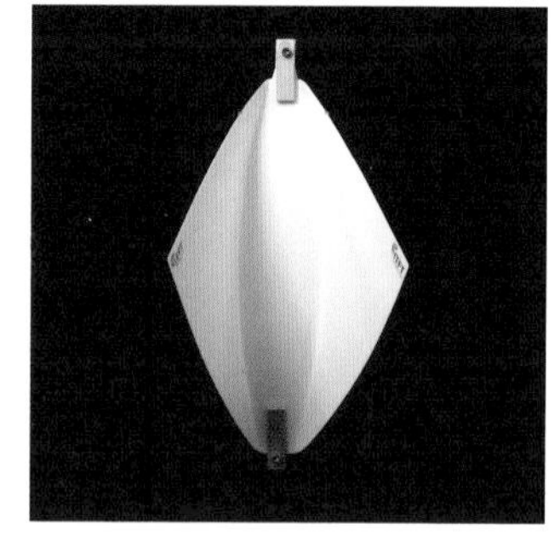

researchers at the University of Stuttgart Institute of Building Structures and Structural Design have developed adaptable envelope components based on deformation principles found in plants. The Flectofin is a fiber-reinforced polymer fabric that provides adaptive shading via elastic deflection (see page 238). **↙ Another technology, Flectofold, interprets the kinetic mechanism of the underwater carnivorous plant *Aldrovanda vesiculosa*, a kind of "living hinge."** This interest in self-folding, articulated textiles has inspired a wave of resilient material surfaces that are simultaneously rigid yet pliable. Trex Lab Technology's Orimetric surface is a tessellated textile that exhibits shock absorption, sound absorption, and collapsibility (see page 242). Designer Oluwaseyi Sosanya's 3D Weaver process fulfills similar objectives, but, in this case, creates thick three-dimensional textiles for protective applications, including body armor and protective sportswear (see page 228).

Next-generation textiles' emphasis on biology is a generative approach, in that it has encouraged consideration of not just living organisms, but also an expansive range of structural and material effects interpreted at diverse scales. For example, the ETFE cladding of the Beijing National Aquatics Center, or "Water Cube," takes inspiration from the morphology of aggregated soap bubbles, effectively scaling a small hydrological phenomenon to the size of a large building. Alternatively, **↖ yet|matilde's Continuous Function adopts the structural logic of thin-shell-engineered structures at the scale of furniture, transforming soft jute fibers into rigid, undulating membranes for seating and storage elements (see page 236).** Albers presaged this trend in her assessment of the blurring of architecture and fabric, stating that the difference "has grown less clearly defined as new methods, affecting both building and weaving, are developing and are adding increasingly to fusion as opposed to linkage."[8] Textiles offer the capacity to assimilate material logics from myriad fields and translate them into inventive manifestations.

---

1 Anni Albers, "The Pliable Plane; Textiles in Architecture," *Perspecta* 4 (Summer 1957): 40.
2 Ibid.
3 Mark Garcia, "Prologue for a History and Theory of Architextiles," *AD: Architextiles* (November/December 2006): 19.
4 Julian Melchiorri profiled by *Deezen*, July 25, 2014, http://www.dezeen.com/2014/07/25/movie-silk-leaf-first-man-made-synthetic-biological-leaf-space-travel/.
5 See accessed January 19, 2016, http://www.technologyreview.com/news/424989/stick-on-electronic-tattoos/.
6 See accessed January 19, 2016, http://www.kaist.edu/_prog/_board/?mode=V&no=10472&code=ed_news&site_dvs_cd=en&menu_dvs_cd=0601&list_typ=B&skey=&sval=&smonth=&site_dvs=&GotoPage=4.
7 See accessed January 19, 2016, http://news.mit.edu/2010/acoustic-fibers-0712.
8 Albers, "Pliable Plane," 36.

THREE-DIMENSIONAL WOVEN TEXTILES

3D Weaver

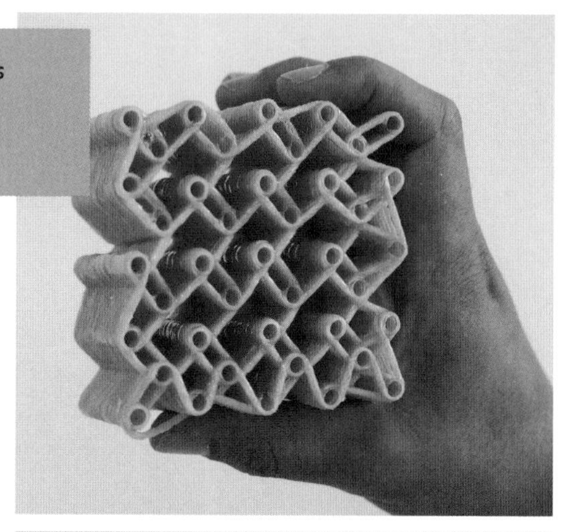

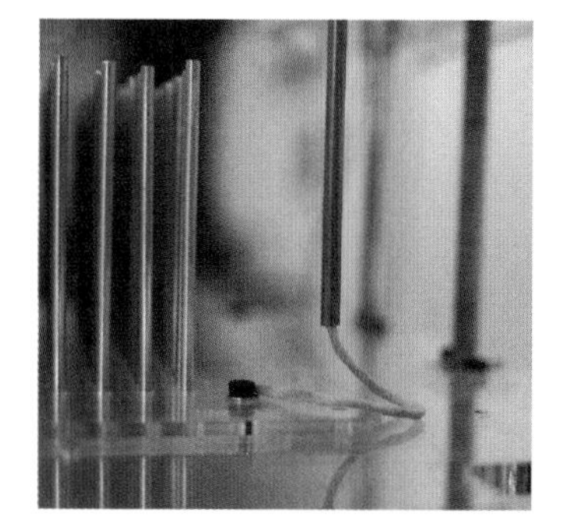

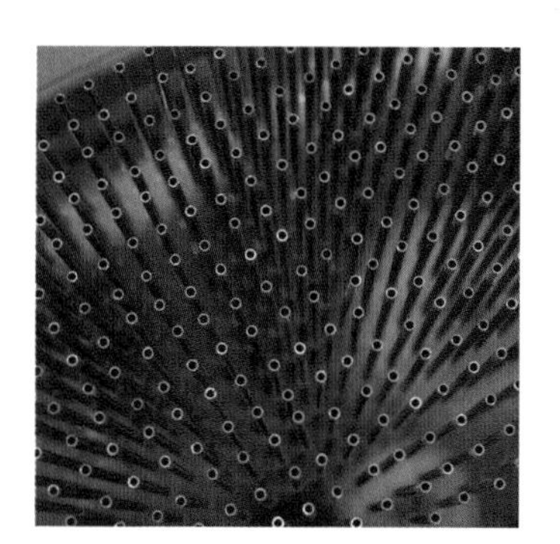

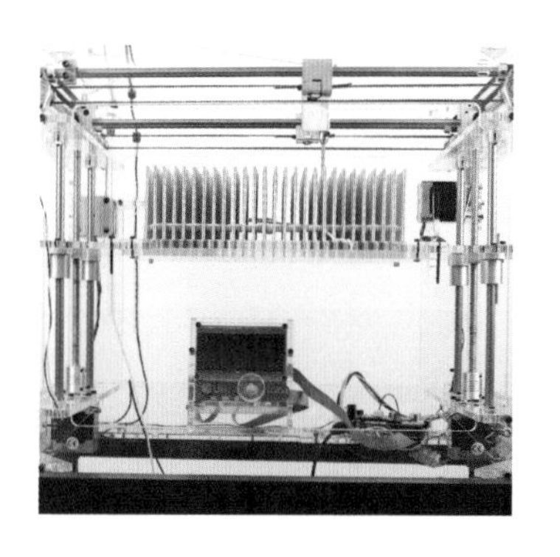

The 3D Weaver is a loom that produces three-dimensional textiles. Developed by designer Oluwaseyi Sosanya, the loom creates resilient textile structures in various patterns using a range of renewable feedstocks, such as paper, wool, and cotton. The machine operates on G-code, a widely used numerical control programming language typically employed in industrial CNC fabrication. The 3D Weaver adopts a structurally robust approach to creating three-dimensional shapes: tubes replace standard warp threads, and the machine pulls the weft threads tightly around the warp elements.

The 3D Weaver can be programmed to produce several patterns. The 0–90 degree design is suitable for taking compressive loads in applications like shoe soles. The Honeycomb pattern, which has a hexagonal structure, is appropriate for impact loads greater than 300 g. The lightweight, flexible ZigZAG weave exhibits auxetic behavior, which means it becomes thicker when elongated. These innovative structured textiles have many potential applications in medical, architecture, vehicle, aerospace, and sportswear industries.

CONTENTS

Cotton, wool, paper, or other organic materials

APPLICATIONS

Footwear, protective sportswear, body armor, compressive building insulation, medical apparatuses

TYPES / SIZES

Three weave patterns: 0–90 degree, Honeycomb, ZigZAG

ENVIRONMENTAL

Makes use of renewable, biocompatible materials; monomaterial composition optimizes recycling

LIMITATIONS

Requires threaded feedstock with moderate tensile strength

FUTURE IMPACT

Biobased soft textiles could replace conventional petroleum products, reducing carbon footprint; potential DIY application of 3D Weaver machines

COMMERCIAL READINESS

CONTACT

Sosa Fresh
Kensington Gore, London SW7 2EU, UK
+44 (0) 7947540107
www.sosafresh.com
ososa@sosafresh.com

MULTIFUNCTIONAL BACTERIAL COMPOSITE

Bio-Fabrication

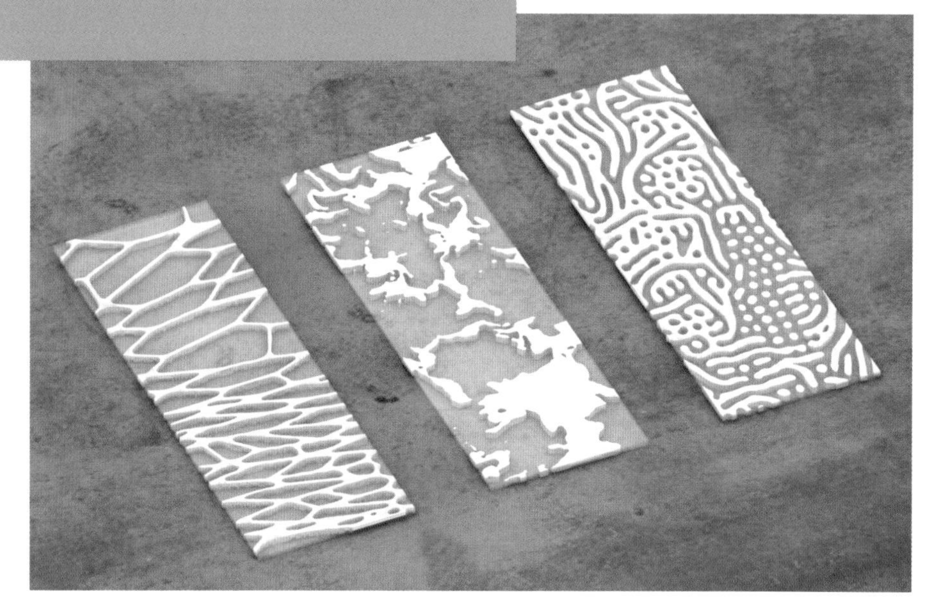

DESIGN VIA EVOLUTIONARY COMPUTATION

DESIGN 00001	DESIGN 00002	DESIGN 00003	DESIGN 00004	DESIGN 00005	DESIGN 00006	DESIGN 00007	DESIGN 00008	DESIGN 00009	DESIGN 00010
DESIGN 01001	DESIGN 01002	DESIGN 01003	DESIGN 01004	DESIGN 01005	DESIGN 01006	DESIGN 01007	DESIGN 01008	DESIGN 01009	DESIGN 01010
DESIGN 02001	DESIGN 02002	DESIGN 02003	DESIGN 02004	DESIGN 02005	DESIGN 02006	DESIGN 02007	DESIGN 02008	DESIGN 02009	DESIGN 02010
DESIGN 03001	DESIGN 03002	DESIGN 03003	DESIGN 03004	DESIGN 03005	DESIGN 03006	DESIGN 03007	DESIGN 03008	DESIGN 03009	DESIGN 03010
DESIGN 04001	DESIGN 04002	DESIGN 04003	DESIGN 04004	DESIGN 04005	DESIGN 04006	DESIGN 04007	DESIGN 04008	DESIGN 04009	DESIGN 04010
DESIGN 05001	DESIGN 05002	DESIGN 05003	DESIGN 05004	DESIGN 05005	DESIGN 05006	DESIGN 05007	DESIGN 05008	DESIGN 05009	DESIGN 05010

Sheets of material are optimized for maximum strength and maximum transparency. Each sheet is rolled into a half-cylinder and tested for structure. Red represents rigid, opaque substance and gray represents flexible, transparent substance.

Bio-Fabrication is a collection of multi-material composites made by bacteria. Recognizing that bacteria can be utilized to fabricate materials with different properties, design firm the Living seeks to combine multiple properties—such as flexibility and rigidity, transparency and opacity—into single composite swatches of material akin to structural textiles. The two-step process begins with the development of new synthetic bacteria strains that produce materials with three different properties: rigidity, flexibility, and intricate patterning. Next, the designers employ the bacteria to grow composite sheets out of sugar. The novel biopolymer textiles exhibit both strength and transparency and are more environmentally friendly than petroleum-derived plastics.

CONTENTS
Synthetic bacteria

APPLICATIONS
Composite-material fabrics, films, and panels for high-performance building envelopes

TYPES / SIZES
Three fundamental material properties: rigidity, flexibility, and intricate patterning

ENVIRONMENTAL
Sugar-based growth medium used instead of petroleum derivations

FUTURE IMPACT
Variegated-property materials for design and building construction created using cultivated combinations of engineered bacteria

COMMERCIAL READINESS

CONTACT
The Living
63 Flushing Avenue, Brooklyn, NY 11205
www.thelivingnewyork.com
life@thelivingnewyork.com

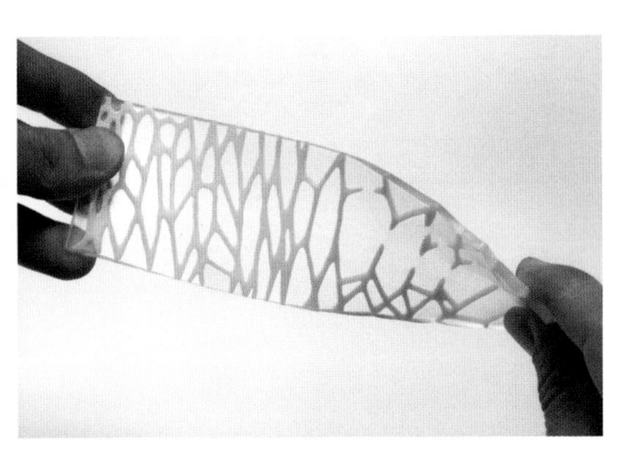

SOUND-MODULATING ARCHITECTURAL TEXTILE SYSTEM

Bloom

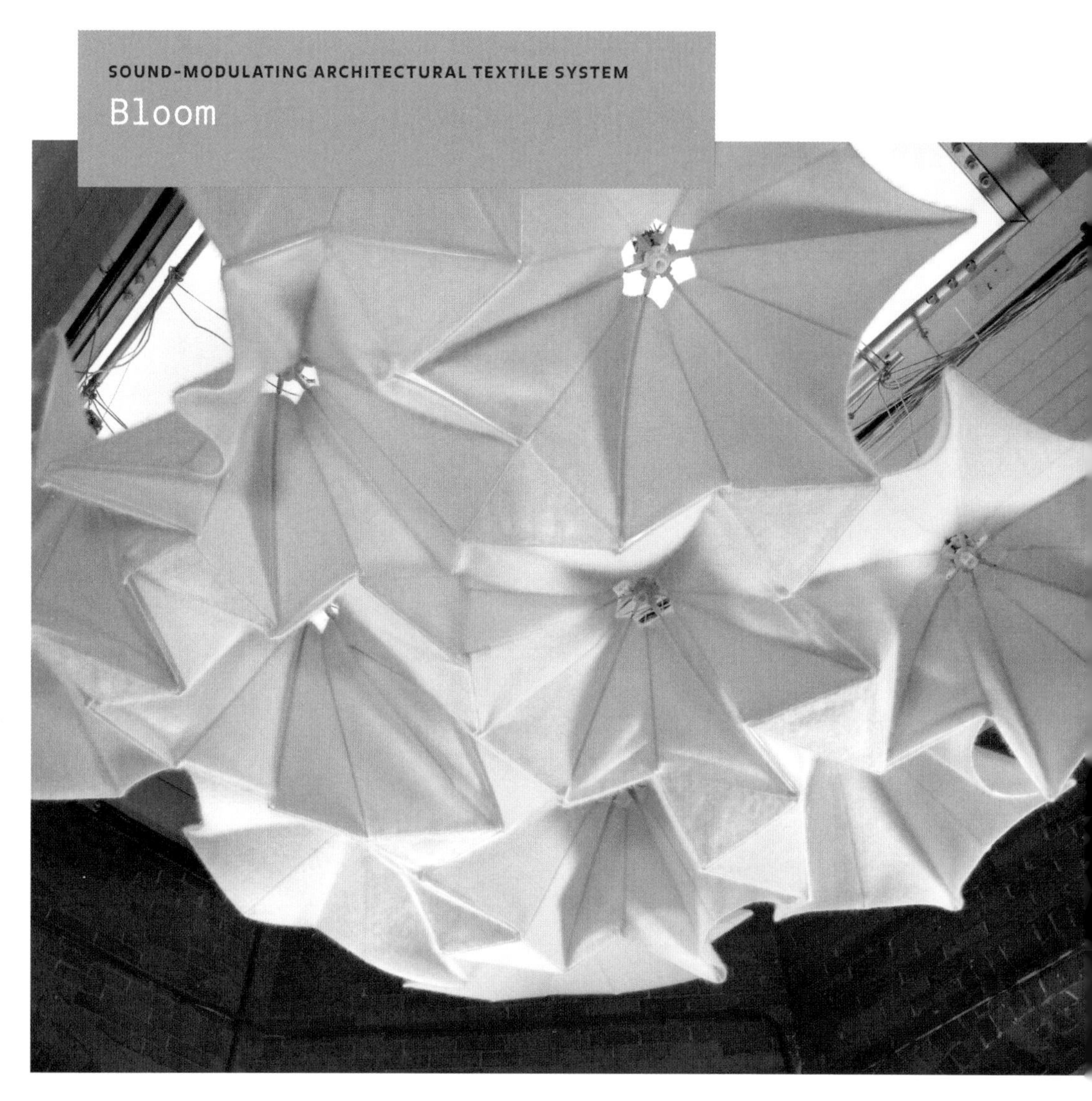

PRODUCT

Bloom is an acoustically tunable textile surface designed for spaces with varying acoustic needs. The shape-changing suspended ceiling is composed of knitted mohair fabric and automatically opens and closes to modulate noise. Bloom's geometry consists of slightly varied triangular shapes, which enhances its capacity to absorb undesired sound waves when open. Alternatively, when users want a live acoustic environment, Bloom collapses to permit increased reverberation.

Envisioned as an interactive acoustic cloud by Yeadon Space Agency in collaboration with Jsssjs Product Design and WickesWerks, Bloom is a highly customizable system in which material, scale, and color may be modified to suit a particular space. The system is ideal for interior environments with demanding acoustic needs, such as conference rooms, restaurants, libraries, auditoriums, and airports.

CONTENTS
Mohair fabric, steel rods, PLA-printed hardware, stepper motors, drivers, microprocessors

APPLICATIONS
Spatial acoustic modulation and interior noise control

TYPES / SIZES
Customizable

LIMITATIONS
For interior use only

FUTURE IMPACT
Ability to tune spaces acoustically with a minimal resource impact

COMMERCIAL READINESS

CONTACT
Yeadon Space Agency
33 Flatbush Avenue, 6th Floor, Brooklyn, NY 11217
www.yeadonspaceagency.com
agent@yeadonspaceagency.com

TRANS-FORMATIONAL

PATTERN-CHANGING TEXTILE

Chromosonic

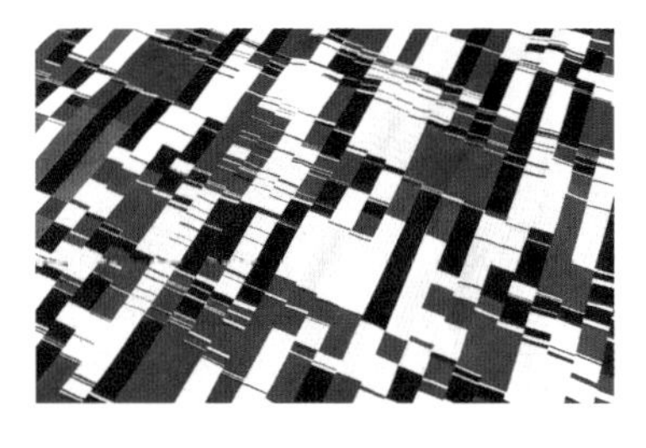

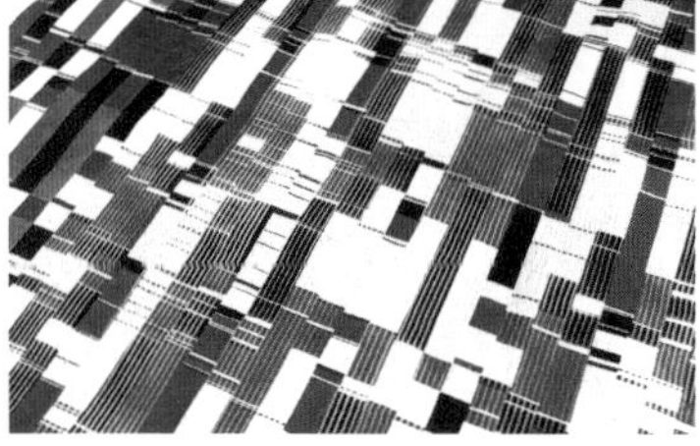

Chromosonic is a programmable electronic, color- and pattern-changing textile interface. The heat-responsive textile's coloration changes dynamically in reaction to processed sound files, which are converted into heat energy, as well as direct contact with the warmth of users' hands.

Designers Judit Eszter Kárpáti and Esteban de la Torre created the interface using linen that is screen-printed with thermochromic dye. An Arduino open-source platform with a 12 V power supply and twenty custom printed circuit boards drive four industrial 24 V DC power supplies. These supplies heat nichrome wires, which are woven into the fabric, transforming the colors of the thermochromic dyes.

Chromosonic investigates how the world of digital media becomes tangible through a textile. The slow-changing fabric responds to ambient environmental impulses and direct user interaction, demonstrating that digital interfaces need not be exclusively defined by backlit glass planes.

CONTENTS

Linen, resistive wire, chromatic pigments, custom electronics

APPLICATIONS

Sound visualization, interactive walls, programmable interfaces, textile structures

TYPES / SIZES

78 $^{3}/_{4}$ × 23 $^{1}/_{2}$" (200 × 60 cm)

LIMITATIONS

For interior use only

FUTURE IMPACT

Increasing the responsive capacity of textiles, use of printed fabric as information display

COMMERCIAL READINESS

CONTACT

EJTech
35 Barsony Street, Miskolc 3531, Hungary
+36 305865352
www.ejtech.cc
we@ejtech.cc

JUTE-COMPOSITE FURNITURE

Continuous Function

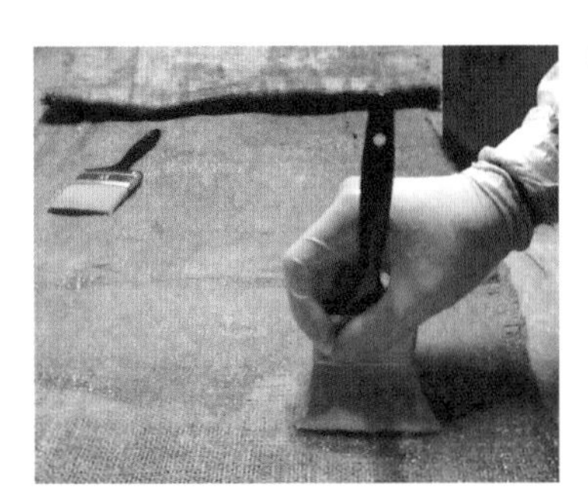

Technology transfer is a potent vehicle for design; it is a productive strategy that involves co-opting the materials or methods from one discipline for another. For example, architects have long held an interest in apparel design, borrowing products and approaches to create architectural textiles. As Torino-based yet|matilde reveals, textiles can also inform the world of furniture design.

Continuous Function transforms soft fabrics into geometric structures suitable for seating, tables, storage, and other furniture. The collection utilizes jute, which is the second most frequently used fiber after cotton, in an unexpected way. To create Continuous Function, the designers fabricate three-dimensional molds upon which they layer multiple sheets of jute fabric. Next, they apply epoxy resin to make the material rigid. Additional jute layers add further structural reinforcement.

Although the coupling of natural fiber and epoxy resin is not optimal from an ecological standpoint, the future substitution of bio-based resin will deliver a satisfyingly biocompatible result. From a material transformation perspective, the metamorphosis of supple cloth into a rigid geometric structure represents an inventive demonstration of technology transfer in design.

CONTENTS

Jute fibers, resin

APPLICATIONS

Seating, tables, storage

TYPES / SIZES

Stool: 19 11/16 × 19 11/16 × 19 11/16"
(50 × 50 × 50 cm), 11 lb. (5 kg)

Table: 39 3/8 × 19 11/16 × 19 11/16"
(100 × 50 × 50 cm), 22 lb. (10 kg)

Double Table: 19 11/16 × 39 3/8 × 19 11/16"
(50 × 100 × 50 cm), 22 lb. (10 kg)

High Table: 39 3/8 × 19 11/16 × 19 11/16"
(100 × 50 × 50 cm), 22 lb. (10 kg)

Long Table: 19 11/16 × 78 3/4 × 19 11/16"
(50 × 200 × 50 cm), 44 lb. (20 kg)

Bookcase: 78 3/4 × 39 3/8 × 19 11/16"
(200 × 100 × 50 cm), 88 lb. (40 kg)

FUTURE IMPACT

Transformation of textiles into structured composites for weight-bearing applications

COMMERCIAL READINESS

●●○○○

CONTACT

yet|matilde
Lungo Dora Firenze 129, Torino 10153, Italy
+39 0110268630
www.yetmatilde.it
yet@matilde.it

FIBER-REINFORCED PLASTIC WITH DIFFERENTIATED STIFFNESS

Flectofin

The Flectofin is a hingeless flapping apparatus inspired by the deformation principle found in the plant Strelitzia reginae, a flowering plant native to South Africa. The plant's valvular pollination mechanism exhibits nonautonomous movement, inspiring Institute of Building Structures and Structural Design researchers to investigate the fundamental underlying principles responsible for the plant's mechanical behavior. The scientists' abstracted model, composed of glass fiber and thermoset resin, reveals a flexible mechanism at the root of the deformation.

Unlike the hinged mechanisms traditionally employed in industry, Flectofin exhibits its elastic deflection without the need for an assemblage of sensitive mechanical elements. The system takes advantage of fiber-reinforced polymers, which have a low stiffness to strength ratio. The glass reinforcing fibers are light transmitting, weather resistant, and inexpensive compared with carbon fibers. Flectofin also features a variation in fabric layers, reducing the amount of material toward the lamella edge to distribute loads over a wide area while minimizing local stress concentrations.

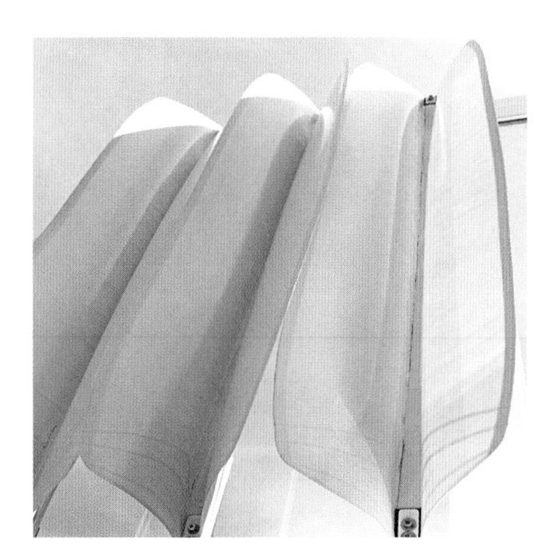

CONTENTS

50 percent glass fiber, 50 percent thermoset resin

APPLICATIONS

Adaptive shading, responsive envelopes

TYPES / SIZES

7 7/8 × 1" – 6' 6 3/4" × 10" (200 × 25 – 2,000 × 250 mm)

ENVIRONMENTAL

Reduces heat gain in buildings, low power consumption during fabrication

TESTS / EXAMINATIONS

DIN 53121: 2008–10 testing of paper and paperboard determination of bending stiffness by the beam method; DIN EN ISO 14125 determination of bending properties; determinations on creep behavior

LIMITATIONS

Maximum length of 6' 6 3/4" (200 cm)

FUTURE IMPACT

Tunable building components for solar shading and view control, flexible systems based on material phenomena rather than mechanical assemblies

COMMERCIAL READINESS

CONTACT

University of Stuttgart
Institute of Building Structures and Structural Design
Kellerstrasse 11, Stuttgart 70174, Germany
+49 (0) 71168583280
www.itke.uni-stuttgart.de
info@itke.uni-stuttgart.de

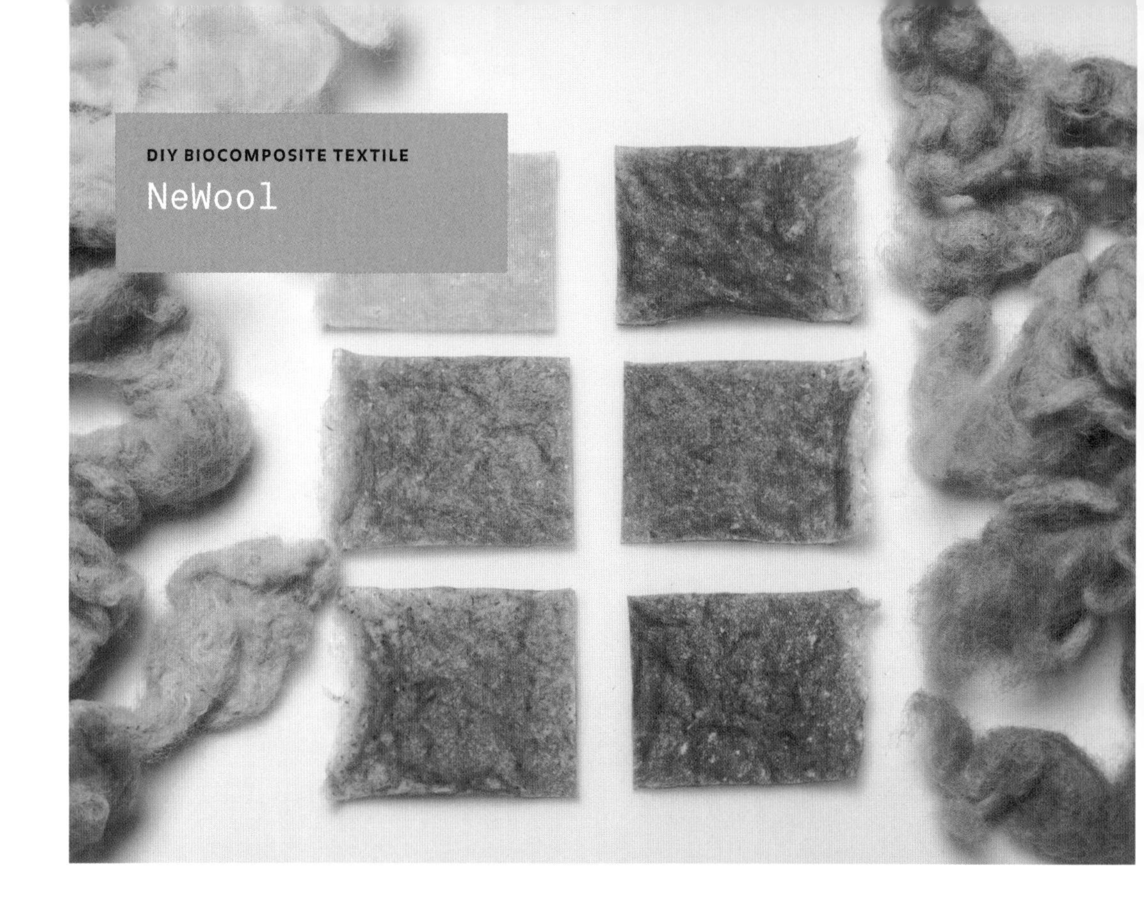

PROCESS

Cook.to.Design is a playful approach to DIY in which anyone can experiment with materials to develop new products. Politecnico di Milano designers Giada Lagorio, Valentina Rognoli, and Marta Rink have used the process to create NeWool, a DIY biocomposite textile. The material consists of repurposed wool fibers, potato starch, and natural pigments. The DIY textile collection includes four different types of fabrics, created using various proportions of wool and potato starch, as well as six color options.

NeWool's inspiration came from concerns about wool waste disposal and interest in optimizing the production chain for wool in Italy. According to the designers, the material represents a more productive intersection between food chemistry, material-based design, and material engineering than traditional manufacturing processes.

CONTENTS

Wool waste, potato starch, natural pigments

APPLICATIONS

Fabric walls, acoustic absorption, rugs or mats, apparel, soft molded objects

TYPES / SIZES

Collection of four materials with different percentages of potato starch and wool waste, six color iterations

ENVIRONMENTAL

100 percent organic ingredients and biocompatible production process, repurposes material waste

LIMITATIONS

Material aesthetics reflect handmade process

FUTURE IMPACT

DIY materials may fill an important niche in industrial ecosystems, making productive use of discarded resources and small-scale, eco-friendly ingredients

COMMERCIAL READINESS

CONTACT

Giada Lagorio
Deymanstraat 14A, Amsterdam 1091 SE, Netherlands
+31 614227113
www.giadalagorio.com
giada.lagorio@gmail.com

RUBBERIZED ORIGAMI TEXTILE

Orimetric

Orimetric is a flexible, structured textile made from geometric patterns derived from origami. Trex Lab Technology has developed a series of material-molding and die-pressing processes that make sharp crease patterns in flat materials, like rubber, leather, and various textiles. The computationally tunable method alters the material properties, making the sheet simultaneously rigid and flexible. Orimetric's valley-and-mountain tessellation pattern imparts materials with excellent shock absorption, sound absorption, and collapsibility down to $^1/_8$" (3 mm), and the maximum ridge height is 2 $^3/_8$" (6 cm).

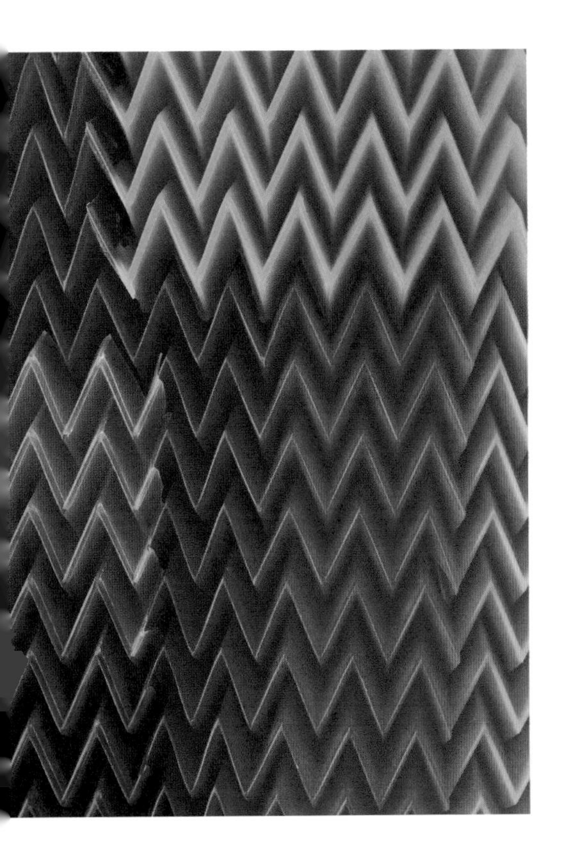

CONTENTS

Rubber, leather, textiles, or other flexible materials

APPLICATIONS

Wall coverings, structured fabrics, curtains, footwear, protective wear, bags, cases

TYPES / SIZES

Various material and color options, 30 × 30" (76 × 76 cm) tile size

ENVIRONMENTAL

Monomaterial construction for easy recyclability

LIMITATIONS

Minimum (compressible) height of $^1/_8$" (3 mm), maximum height of 2 $^3/_8$" (6 cm)

FUTURE IMPACT

Ability to provide soft materials with structure and elasticity to create resilient material systems

COMMERCIAL READINESS

CONTACT

Trex Lab Technology
160 Scholes Street, Brooklyn, NY 11206
www.trex-lab.com
info@trex-lab.com

3D TEXTILE-CLAD FURNISHINGS

Poli

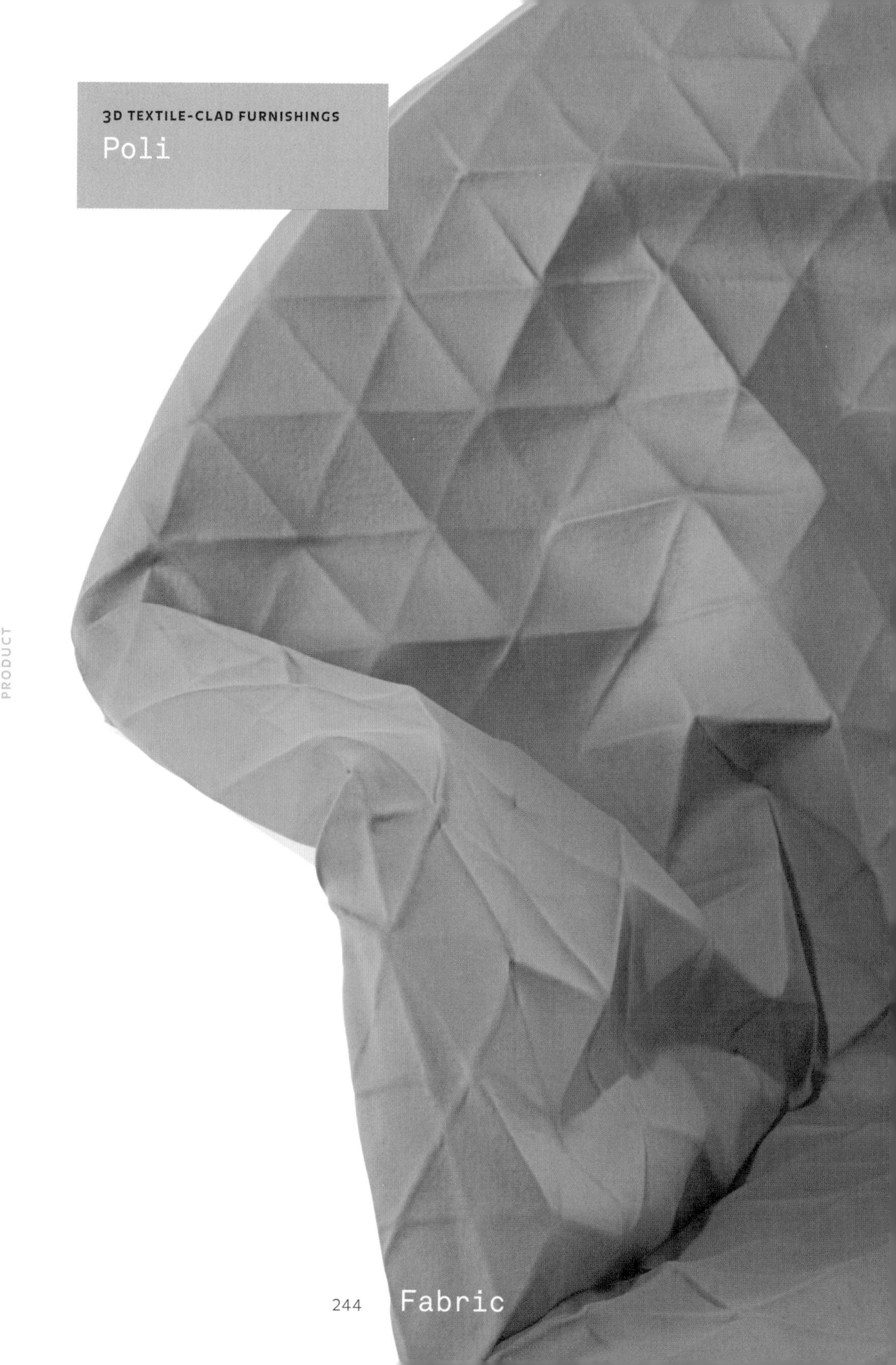

Poli is a lounge chair and ottoman set designed by Producks Design Studio in collaboration with textile designer Mika Barr. Each piece of furniture consists of a single multidimensional textile that covers an aluminum and steel frame. The Poli textile exhibits a soft internal structure and naturally conforms to various adjacent surfaces, such as the seat, back, and armrests of a chair. The use of a single expanse of textile avoids the need for stitching, allowing the interior foam to be directly molded to the fabric.

CONTENTS
Microfiber textile, polyurethane, aluminum, steel

APPLICATIONS
Furniture, soft structured surfaces

TYPES / SIZES
Lounge chair, ottoman

ENVIRONMENTAL
Reduces material waste during fabrication

FUTURE IMPACT
Creation of structured geometric surfaces using soft textiles

COMMERCIAL READINESS

CONTACT
Producks Design Studio with Mika Barr
58 Gordon Street, Ramat Hasharon 47263, Israel
+972 (0) 523632611
www.pro-ducks.com
info@pro-ducks.com

MATERIAL

SYNTHESIZED SPIDER SILK

Qmonos

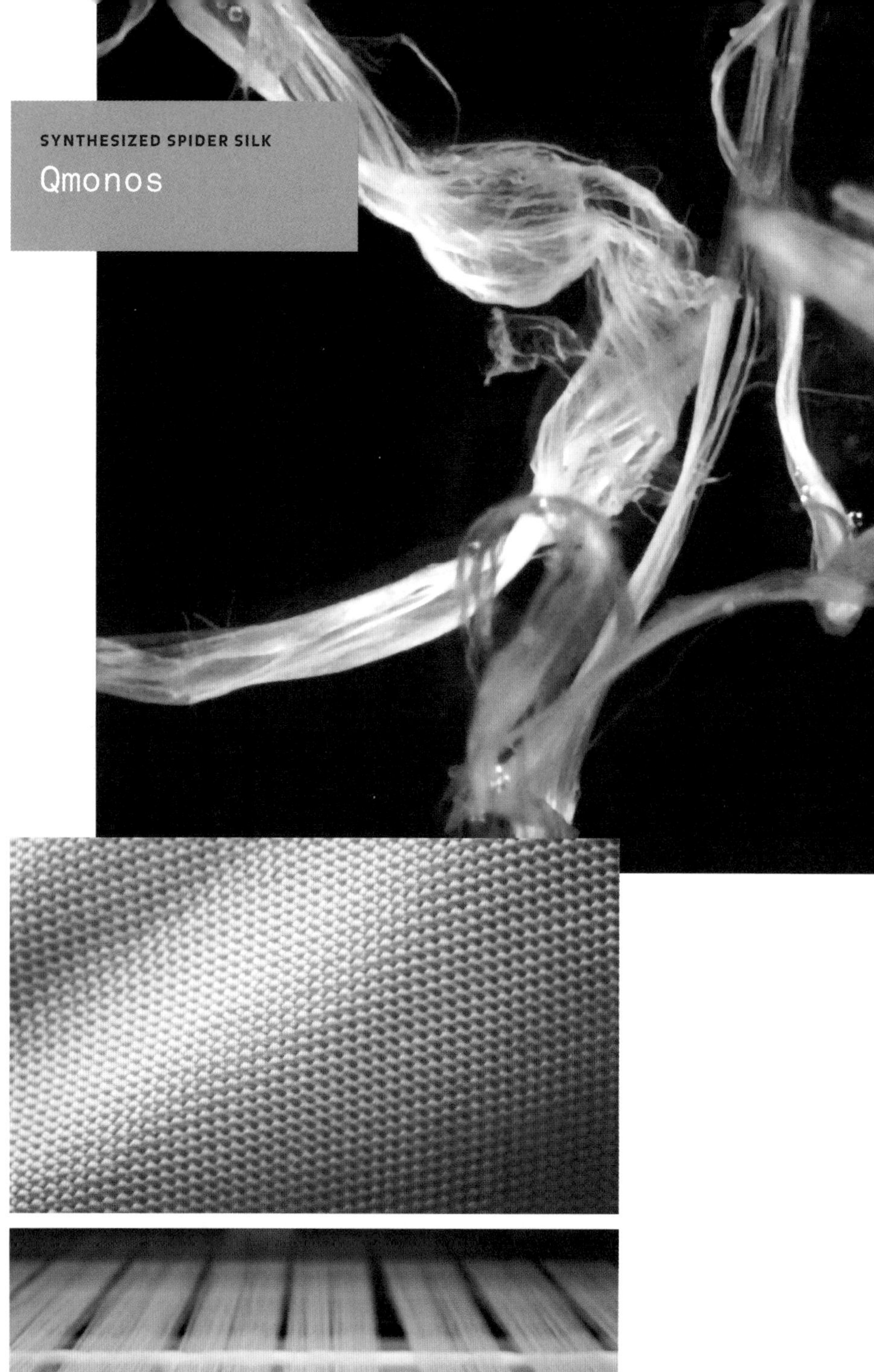

Spider silk has long fascinated scientists because of its incredible strength and formation process. Stronger than steel by weight and more durable than Kevlar, the material also has amazing elasticity. Moreover, it is created biochemically, as opposed to requiring the addition of heat. Nexia Biotechnologies developed an artificial spider silk in the late 1990s called BioSteel, which involved growing spider silk protein in transgenic goat milk. The company has since folded; however, Japan-based Spiber has continued the effort to synthesize fibroin, which is the protein that imparts dragline silk with its compelling characteristics.

By decoding the fibroin-producing gene, Spiber has avoided using goat milk—instead producing the protein with a marriage of bioengineered bacteria and recombinant DNA. The company processes the protein as a dry powder from which threads are manufactured by extruding the powder through thin, hollow needles. Spiber calls this fiber Qmonos, based on *kumonosu*, the Japanese word for spiderweb.

In late 2015, Spiber announced a collaboration with the North Face to produce a jacket for extreme environments, like the South Pole. Claimed to be the first article of clothing made from synthetic proteins, the so-called Moon Parka features an outer shell made from woven Qmonos, which is dyed with a soft golden hue reminiscent of the natural silk color of the golden orb spider. This shell not only outperforms conventional petroleum-derived fibers, but also is entirely biodegradable.

CONTENTS

Bioengineered fibroin

APPLICATIONS

High-performance textiles, apparel, medical devices, automotive and architectural fabrics

TYPES / SIZES

Over six hundred proteins and twenty types of amino acids available for recombinant processing, various color dyes available

ENVIRONMENTAL

Manufactured using an ecologically responsible process, biodegradable

LIMITATIONS

Requires a ten-day period between gene synthesis and test spinning

FUTURE IMPACT

Fabrication of ultrastrong, ecologically sensitive products based on natural proteins

COMMERCIAL READINESS

CONTACT

Spiber Inc.
234-1 Mizukami Kakuganji, Tsuruoka, Yamagata 997-0052, Japan
www.spiber.jp
admin@spiber.jp

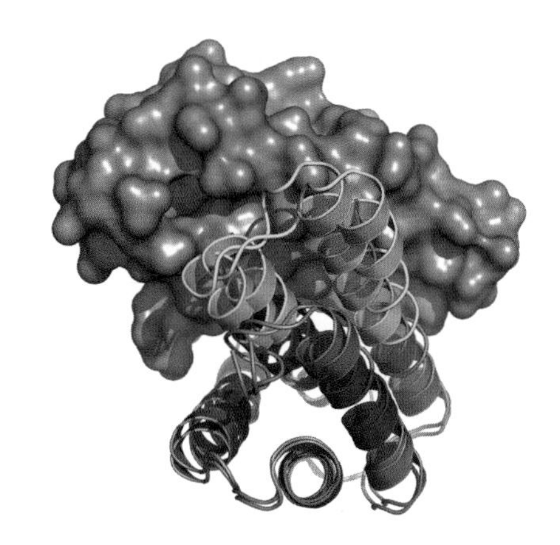

VAPOR-RESPONSIVE ARTIFICIAL SKIN

Self-Flexing Membrane

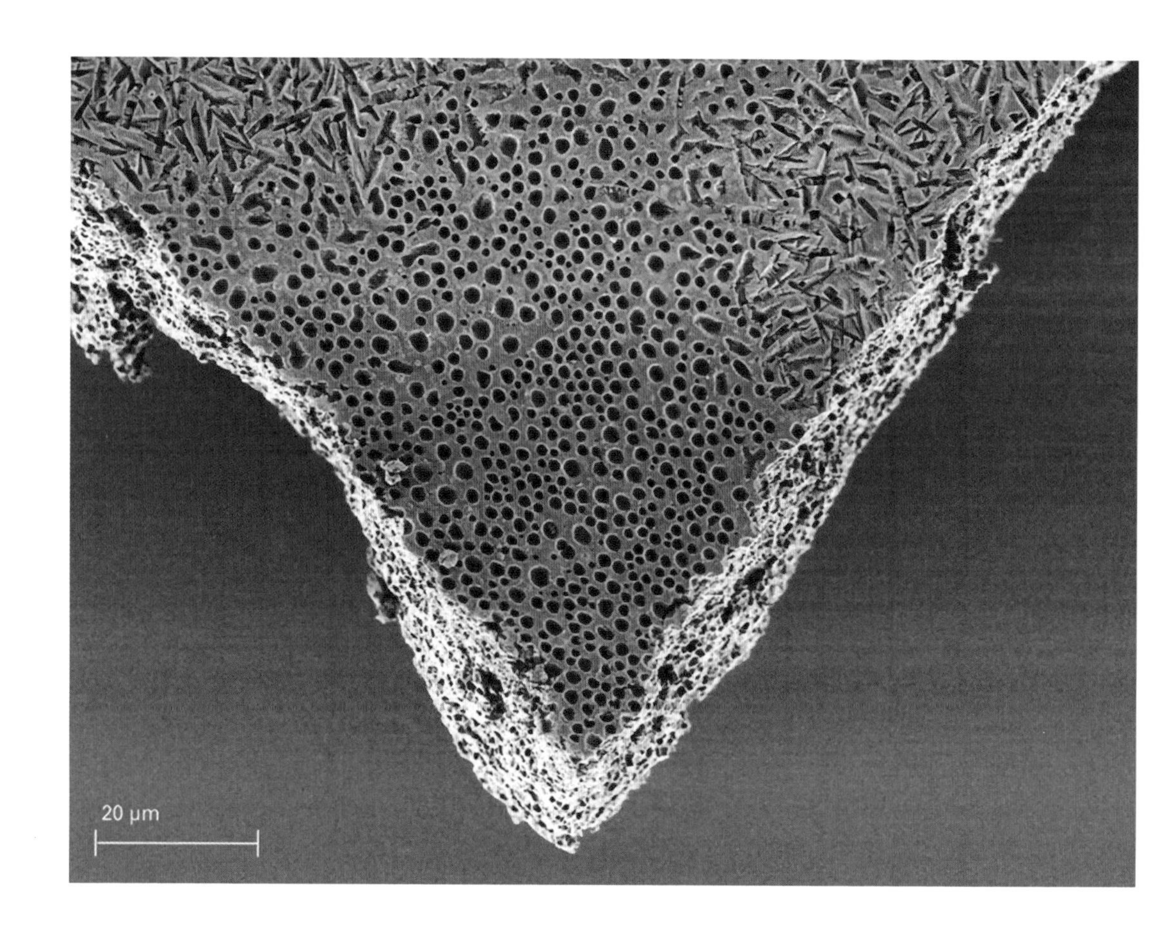

Self-Flexing Membrane is a synthetic skin developed by researchers at the Max Planck Institute of Colloids and Interfaces in Potsdam, Germany. The membrane is extremely sensitive to solvent vapors and quickly curls in the presence of acetone and other organic solvents. The artificial skin's response mechanism is reminiscent of biological examples such as the ice plant seed capsule or Venus flytrap. The reactive foil features layered functionality: the top surface is rigid while the bottom can become soft, thus allowing directional curling. Furthermore, the membrane is perforated with tiny pores that enable the film to detect vapor rapidly. The Self-Flexing Membrane has a faster reaction time than other known actuators and is appropriate for use in sensors as well as artificial robotic skin and musculature.

CONTENTS

Ionic polymer

APPLICATIONS

Autonomic sensors, artificial robotic skin and muscles

TYPES / SIZES

Vary

ENVIRONMENTAL

Passive detection and communication of environmental toxins, efficient use of resources

TESTS / EXAMINATIONS

Solvent actuation, light actuation

LIMITATIONS

Actuator loses reaction speed at extreme temperatures

FUTURE IMPACT

Development of smart, biomimetic actuators that outperform conventional technologies; soft, resilient materials increasingly applied in lieu of rigid mechanical devices

COMMERCIAL READINESS

CONTACT

Max Planck Institute of Colloids and Interfaces
Potsdam-Golm Science Park,
Am Mühlenberg 1 OT Golm,
Potsdam 14476, Germany
+49 8921080
www.mpikg.mpg.de/en
info@mpikg.mpg.de

COLOR-CHANGING FIBERS FOR WOUND HEALING

Smart Bandage

The smart bandage improves wound supervision by communicating healing problems, such as infection. Developed by Australia's Commonwealth Scientific and Industrial Research Organisation, the dressing consists of a fiber with an internal sensory layer of thermochromic liquid crystalline. When illuminated by white light, the reflected color of the crystals changes, enhancing the contrast of the filamentary core. The fiber color gradient operates within a range of 77 to 113 °F (25 to 45 °C) and changes hue with temperature shifts as small as 0.9 °F (0.5 °C). The bandage can thus indicate the development of an infection that would otherwise present few visual cues in its formative stages. This accurate monitoring capability ensures that wound dressings need not be disturbed unless a problem arises.

CONTENTS
Thermochromic fiber

APPLICATIONS
Wound management, biological monitoring

TYPES / SIZES
Variable sizes

FUTURE IMPACT
Enhanced diagnosis of biological health, accelerated wound healing and recovery from illness

COMMERCIAL READINESS

CONTACT
Commonwealth Scientific and Industrial Research Organisation
PO Box 225, Dickson 2602, Australia
+61 395452176
www.csiro.au
enquiries@csiro.au

9

Light

> Light as Material

> Light is immaterial, yet the ways
> in which we generate light involve
> matter. Lightbulb filaments, laser
> pointers, fireflies, and stars
> all employ material as a vehicle for
> electrical charges to emit light.[1]
> Throughout the ages humankind has
> recognized this material basis
> for light – in the flames produced
> by a brush fire, the burning magma
> of a lava flow, or the soft radi-
> ance of a glowworm's abdomen.
> Humans learned to make and control
> fire as early as one million years
> ago (as *Homo erectus*), a significant
> milestone in our evolution.[2] This
> physical connection to light, a
> source of illumination, warmth,
> safety, and food, undoubtedly left
> its mark on our developing psyche.

Until the advent of electric lighting, light imparted substantial multisensory effects. As historian Jane Brox describes, "Human light has its sounds—of a match struck and a candle flame muttering in a draft, of a stopcock turning and a gas jet hissing to life or hoarsely damping itself out." [3]

Electric illumination, or "light without fire," enabled greater control over light but at the same time made it more abstract.[4] The systematic deployment of electric systems brought about a new lexicon of watts, volts, and filaments. Rapid advances in incandescent technology and the expansion of electricity distribution into rural areas transformed the nocturnal landscape. In 1922 General Electric scientist Charles Steinmetz boasted, "Today we are producing . . . sixty-eight times as much light as we could produce with the lights in use fifteen years ago." [5] The "cold light" of fluorescent technology, developed in the mid-1930s at the General Electric Laboratories, made electric light even more abstract.[6] By midcentury, fluorescent lighting had proliferated widely in public buildings and office spaces throughout the United States, encompassing over half the interior lighting in the country.[7] *Cold* was a suitable term: early fluorescent lighting, which was characterized by poor color rendering, imparted spaces with a pallid, undifferentiated light. Despite this undesirable quality, however, fluorescence came to define the character of mid- to late twentieth-century interior environments, particularly in nondomestic settings.

Lighting is currently experiencing another radical transformation. After decades of uniformity and abstraction, light is once again becoming palpable and personal. Building occupants desire interaction and control, and current lighting technology enables an advanced level of management. Such interactivity is an ever-complex challenge given that natural daylighting has become a much more important and welcome factor in luminous programs. Sophisticated technologies—such as automated interior dimming connected to external building shades, tunable LED fixtures with a high color rendering index, and mobile applications that adjust personal lighting settings based on established preferences—are increasingly common. Light is becoming more manipulable, responsive, environment aware, and integrated into designed objects and environments.

Light is now incorporated into the design of furniture, horizontal and vertical surfaces, mechanical systems, and even clothing. **↙ NONdesigns' Common Desk is a collaborative work surface that discharges light from below through open divider slots (see page 264).** Unlike in typical work spaces, where lighting is provided with discrete fixtures, the tabletop design cleverly integrates illumination into the structure of file organization. In interiors, ceilings have typically been the platform for electric light. However, UK start-up Lomox is developing a light-emitting wallpaper using low-energy OLED technology.[8] Floors are occasionally also luminous terrain, as seen in NighTec light-emitting pavers.[9] Even windows may now be electric light sources, as demonstrated by a switchable OLED technology developed by BASF and Philips, with initial applications in car roofs.[10] Building services are also fair game: Mexico City–based design outfit Hierve ingeniously combines light and plumbing in its Hydroelectric Lamp, which is powered by the pressure exerted by running water.[11] Furthermore, apparel and personal accessories are now fertile territories for experimentation, as seen in the GER color-changing Mood Sweater or the Neclumi projection-based interactive necklace.[12]

Increasingly, living organisms are being utilized as vehicles for lighting design. The Biobulb developed at the University of Wisconsin–Madison is a microbe-powered light source (see page 260). A combination of bioluminescent, genetically modified *E. coli* bacteria, algae, and predatory protists form a self-sustaining, continuously illuminating ecosystem in a bulb. **↓ The Starlight Avatar is the world's first autoluminescent plant, genetically engineered by molecular biologist Alexander Krichevsky to emit light (see page 262).** San Francisco–based inventor Antony Evans is developing a scaled-up version: a glowing tree that can function as a zero-energy streetlight.[13] In the near future, landscapes will be designed based on nocturnal luminous programs for low-level nighttime visibility, wayfinding, and ornamentation.

Interactivity is also on the rise. As lighting technology becomes more personal, it also becomes more haptic. The Ball Wall is a solar-charged surface composed of polymer spheres that photoluminesce in the dark, and users can effectively "draw" by rotating them in place (see page 258). Studio Roosegaarde's Crystal project is a collection of hundreds of translucent salt crystals with internal LEDs and sensors (see page 266). Charged by a magnetic

surface, the crystals brighten when touched and manipulated. **↑ Swing Time, designed by Höweler and Yoon Architecture, is a transitory, illuminated playscape that activates social activity in otherwise unused urban spaces (see page 276).** Meanwhile, the interactive LED wall called Mnemosyne is designed to enhance personal safety within poorly lit areas of dangerous neighborhoods (see page 272). The most immersive example is Waterlicht, or "water light," which is a wave-simulating light projection that transforms extensive outdoor areas into chimerical seascapes (see page 278).

These significant changes in next-generation lighting demonstrate its treatment as a material that we can manipulate, instead of an abstract force beyond our control. Light artist James Turrell describes his installations in this way, in comparison with traditional art: "There is a rich tradition of painting of work about light, but it is not light—it is the record of seeing. My material is light, and it is responsive to your seeing." [14] As energy concerns come to dominate discussions about lighting design, we must not forget the essential objective of enhancing vision. Naturally, designers must create lighting strategies that make optimal use of resources—and fortunately, lighting technology continues to make positive strides in this regard. However, designers must not forget the fundamental role of light to shape experience, and to provoke delight and wonder. "It's about perception," declares Turrell. "For me, it's using light as a material to influence or affect the medium of perception. I feel that I want to use light as this wonderful and magic elixir that we drink as Vitamin D through the skin—and I mean, we are literally light-eaters—to then affect the way that we see." [15]

---

1 Ian A. Walmsley, *Light: A Very Short Introduction* (Oxford: Oxford University Press, 2015), 67.
2 See accessed January 19, 2016, http://news.discovery.com/history/archaeology/human-ancestor-fire-120402.htm.
3 Jane Brox, *Brilliant: The Evolution of Artificial Light* (Boston: Houghton Mifflin Harcourt, 2010), 93.
4 Ibid.
5 Charles Steinmetz quoted in "Scientists Racing to Find Cold Light," *New York Times*, April 24, 1922.
6 Brox, *Brilliant*, 209.
7 Ibid., 213.
8 See accessed January 19, 2016, http://news.bbc.co.uk/2/hi/uk_news/wales/8434705.stm.
9 See accessed January 19, 2016, http://www.aggregate.com/products-and-services/commercial-hard-landscaping/paving/block-paving/nightec/.
10 See accessed January 19, 2016, http://phys.org/news/2012-01-basf-philips-oled-transparent-car.html.
11 See accessed January 19, 2016, http://www.en.hiervé.com/projects/hydroelectric-lamp/.
12 See accessed January 19, 2016, http://sensoree.com/artifacts/ger-mood-sweater/; and accessed January 19, 2016, http://www.neclumi.com.
13 See accessed January 19, 2016, http://www.smithsonianmag.com/innovation/creating-a-new-kind-of-night-light-glow-in-the-dark-trees-9600277/.
14 James Turrell, *Occluded Front* (Los Angeles: Fellows of Contemporary Art, 1985), 43.
15 James Turrell interviewed by Michael Govan, *Interview*, June 30, 2011, http://www.interviewmagazine.com/art/james-turell.

LIGHT-HARVESTING DRAWING SURFACE

Ball Wall

The Ball Wall is an interactive light surface that allows visitors to map gestures by spinning an array of low-friction balls treated with a luminescent coating and mounted on a wood and metal frame. Designers Paul Bird, Conor Oberlander, and Kevin Remy developed the interface at Rhode Island School of Design as a facade application for the 2014 Solar Decathlon Europe.

The user manipulates Ball Wall by moving a hand across hundreds of polypropylene spheres. The orbs are internally coated with photoluminescent paint and set in a hexagonal formation. The system allows a user to draw a picture by treating each sphere as a pixel. The Ball Wall is infinitely scalable, and it is automatically reset by the sun every day. The system may also be configured to fit nonplanar surfaces.

CONTENTS

Acrylic, polypropylene, photoluminescent paint, walnut

APPLICATIONS

Interactive wall installation, visual display

TYPES / SIZES

Hexagonal modular components measuring 11" (28 cm) diagonally

ENVIRONMENTAL

Illumination requires no electricity if only using sunlight

LIMITATIONS

Requires solar exposure or a high-intensity interior light source

FUTURE IMPACT

User-manipulable wall surfaces, interactive visual communications, solar-harvesting building envelopes

COMMERCIAL READINESS

CONTACT

Rhode Island School of Design
2 College Street, Providence, RI 02903
401-454-6100
www.risd.edu

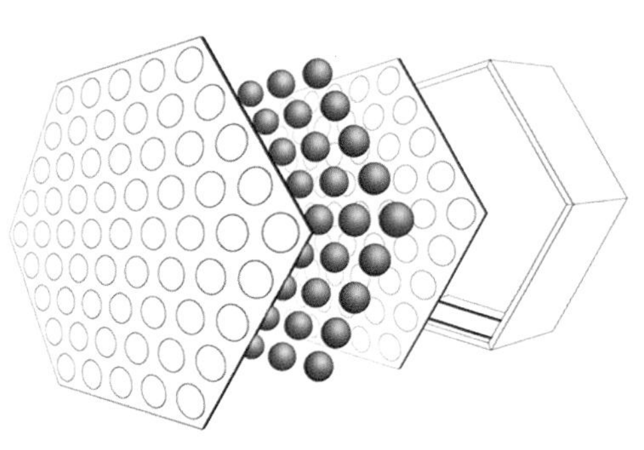

MICROBE-POWERED LIGHT

Biobulb

PRODUCT

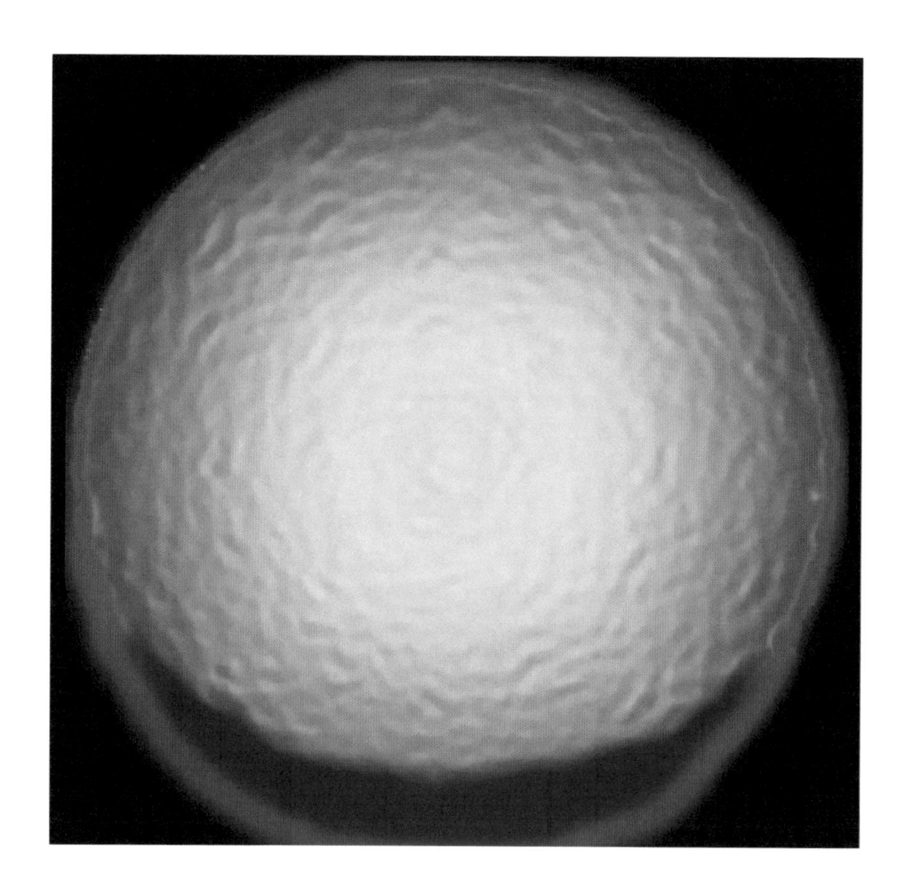

Increased interest in the capability of bioluminescent organisms, which emit light by way of chemical reaction, has inspired some scientists to propose methods for harnessing this phenomenon in designed products. A multidisciplinary team of researchers at the Wisconsin Institute for Discovery is developing a self-contained light source that is powered by light-emitting bacteria.

The Biobulb contains E. coli bacteria that have been genetically modified to emit light. The bulb also includes additional organisms required to manage and sustain its light-producing function, such as algae to harvest energy from sunlight, and predatory protists (eukaryotic organisms) to limit bacteria growth and recycle nutrients. Remarkably, this self-enclosed ecosystem is designed to run continuously on sunlight and managed waste; however, in reality, the bulb life is uncertain, due to the delicate balance of the microorganisms and the likely need for occasional recharging. Nevertheless, the Biobulb represents a promising step toward creating a net-zero light source without using conventional renewable energy.

CONTENTS
Genetically modified *E. coli*, algae, predatory protists, growth media, glass vessel

APPLICATIONS
Low-level illumination

TYPES / SIZES
Vary

ENVIRONMENTAL
No electricity required

LIMITATIONS
Requires frequent and regular maintenance, low-intensity light source

FUTURE IMPACT
Cultivation of live bioengineered microorganisms in the service of practical functions like lighting, transition toward active partnership model with natural systems

COMMERCIAL READINESS
● ○ ○ ○ ○

CONTACT
Wisconsin Institute for Discovery
University of Wisconsin–Madison
330 North Orchard Street, Madison, WI 53715
608-316-4300
www.wid.wisc.edu
info@wid.wisc.edu

UNDERLIT WORK SURFACE

Common Desk

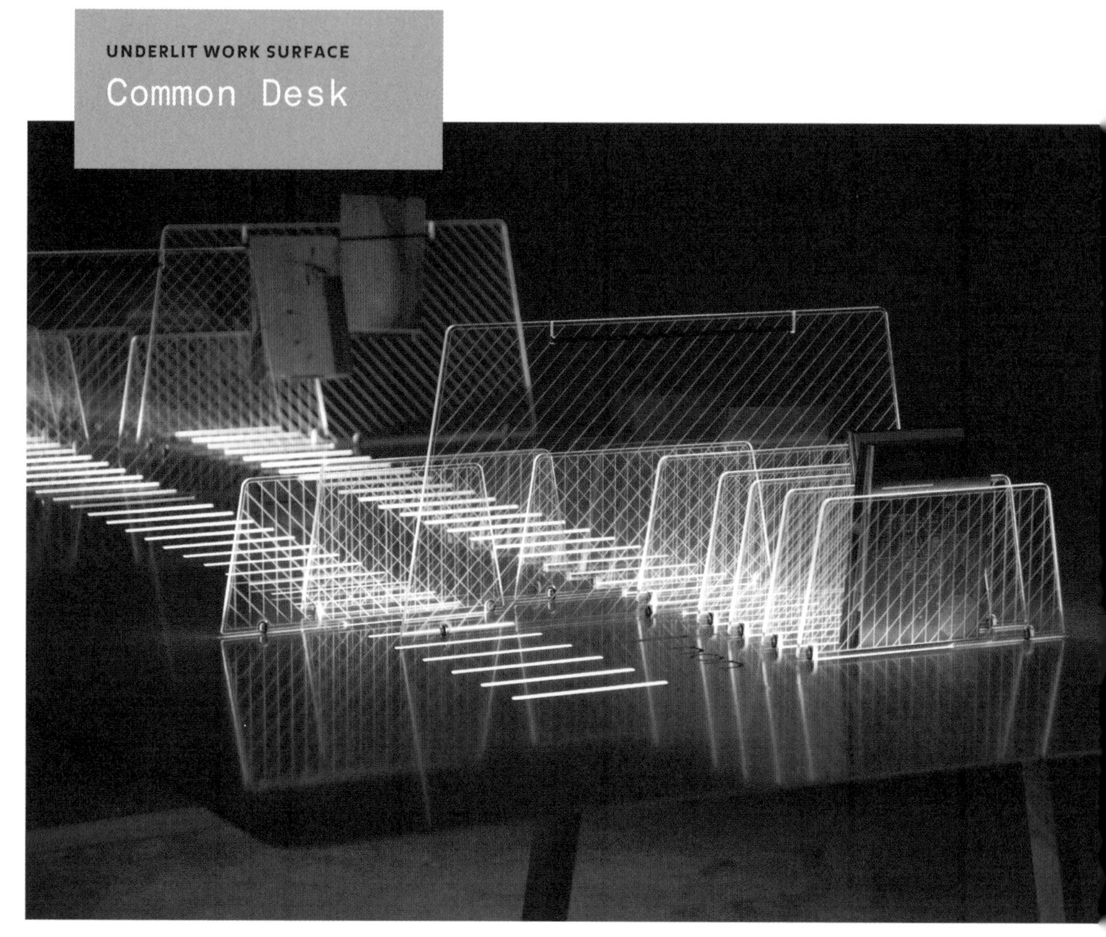

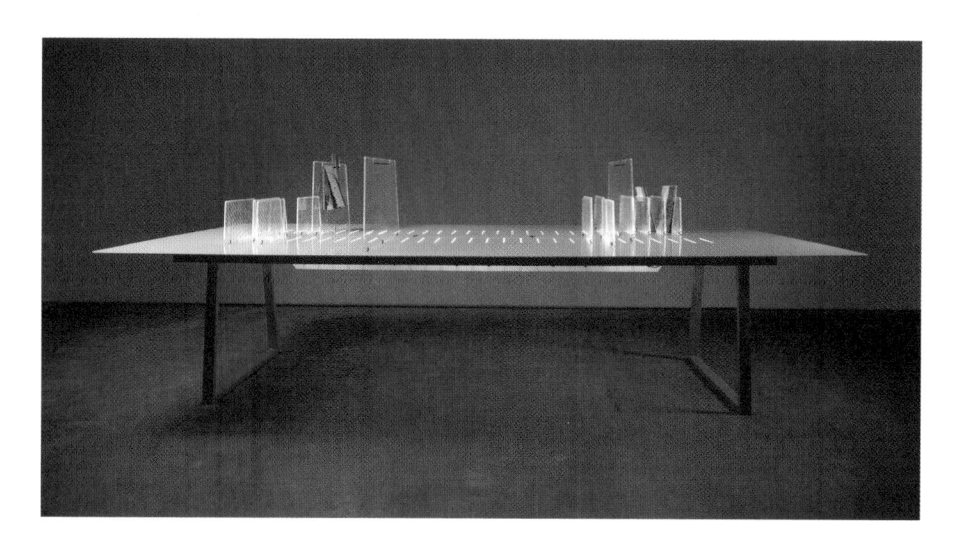

PRODUCT

Pasadena, California–based NONdesigns has created the Common Desk, a surface with an unexpected approach to illumination. The collaborative work space is perforated with a series of linear slots that hold acrylic panels, which function as movable dividers and bookends. From within these slots, the Common Desk glows, providing ample lighting for anyone working late.

The desk comes with three large and twelve small acrylic panels. Each panel channels LED light—the source of which is located under the desk surface—via glowing edges, collectively imparting a diffuse ambiance. A central trough also allows for cable and power supply storage. An imaginative approach to organization and dividing workstations, Common fuses the structure of file management with the emission of light—a novel way to illuminate work.

CONTENTS

Aluminum, acrylic, LED lighting

APPLICATIONS

Work surface, table, light source

TYPES / SIZES

Three large acrylic panels, twelve small acrylic panels, Pantone colors available upon request, desk size 2' 4 1/4" × 9' 11 1/2" × 4' 11 1/2" (152 × 277 × 72 cm)

LIMITATIONS

High-visibility tasks will require additional illumination in a dark space

FUTURE IMPACT

Innovative integration of lighting with other elements of the designed environment, such as furniture, objects, and surfaces

COMMERCIAL READINESS

CONTACT

NONdesigns LLC
299 North Altadena Drive, Unit 120,
Pasadena, CA 91107
626-696-3081
www.nondesigns.com
info@nondesigns.com

INTERACTIVE LIGHT STONES

Crystal

PRODUCT

Crystal is a collection of hundreds of crystals of light that brighten when touched. Designer Daan Roosegaarde developed Crystal to enhance the functionality of the floor as a surface, inviting participants to interact with the horizontally placed illuminated objects to communicate their stories at what he calls a digital campfire.

Each salt crystal has embedded LEDs that are wirelessly powered via a black magnetic floor. Once visitors start moving, adding, or sharing the objects, the crystals' illumination fluctuates in intensity. According to the designer, the lighting behavior of the crystals later transforms from an excited to a resting state, keeping visitors curious and engaged.

CONTENTS

Translucent salt crystals, LEDs, power mat

APPLICATIONS

Interactive artwork, lighting, landscaping

TYPES / SIZES

Crystal and Crystal-S types, colors and sizes vary

ENVIRONMENTAL

Low-power lighting

TESTS / EXAMINATIONS

Public installations in Amsterdam; Paris; Moscow; and Eindhoven, Netherlands

LIMITATIONS

Requires contact with power mat to charge

FUTURE IMPACT

Enhancement of public spaces with interactive light installations, promoting engagement with the landscape via physical interactions

COMMERCIAL READINESS

CONTACT

Studio Roosegaarde
Vierhavensstraat 52–54,
Rotterdam 3029 BG, Netherlands
+31 103070909
mail@studioroosegaarde.net
www.studioroosegaarde.net

VOLUMETRIC LIGHT DEPOSITION

Light Printing

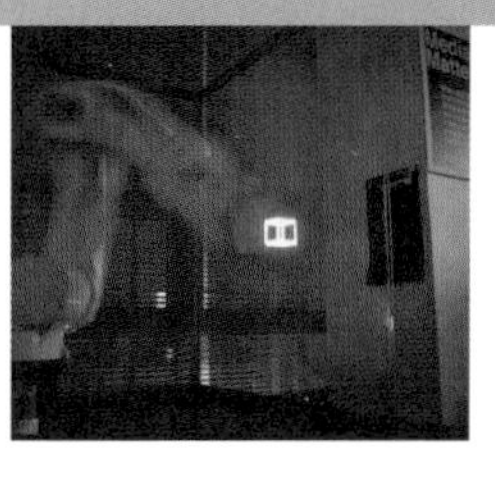

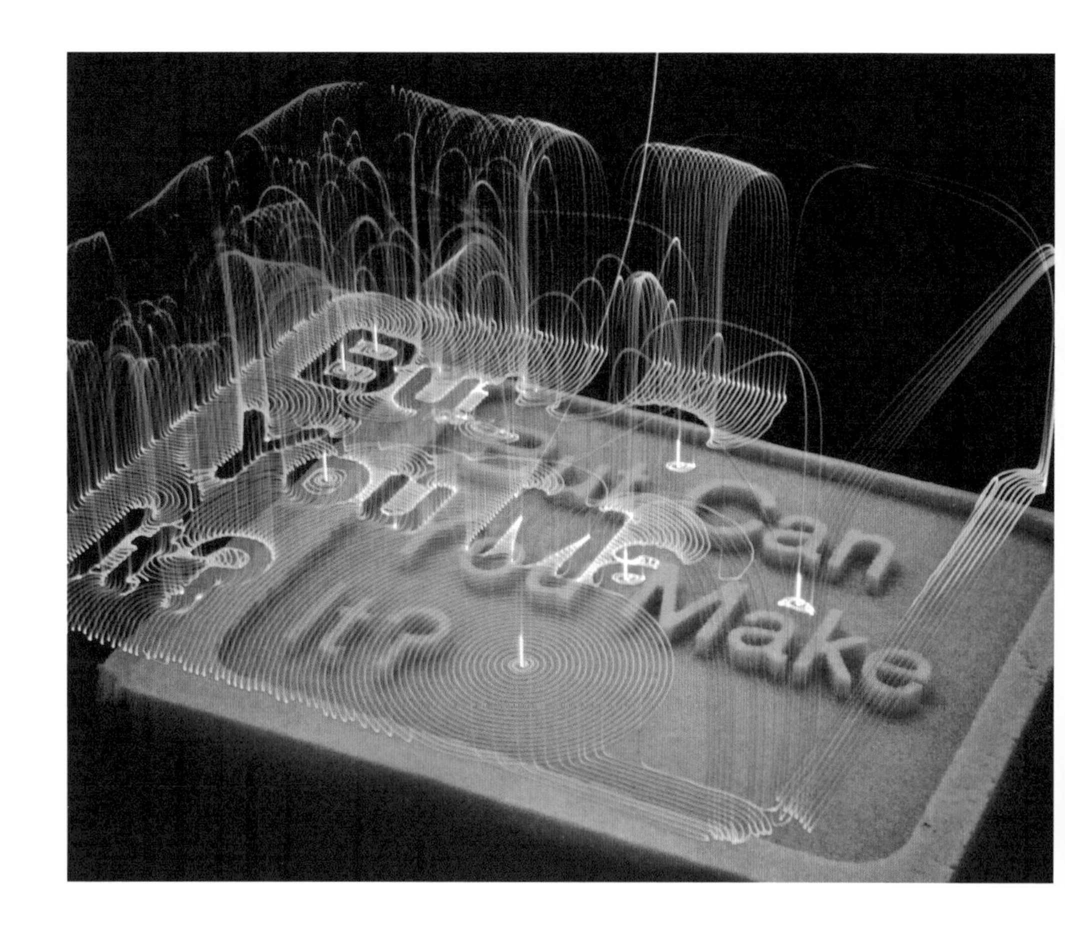

Demonstrating that matter is not the only printable feedstock, Light Printing is a method to create light volumes in space. A robotic arm drives a light source via prescribed movements in front of a long-exposure camera, which records the scene as a volumetric light composition. This light painting technique can also be reversed, in which case the arm controls a camera that captures contextual illumination. Sensors capable of recording invisible radiation—such as magnetic fields, radio waves, or heat currents—can also be employed, enabling sophisticated real-time environmental scanning.

CONTENTS

Robotic arm, light source, camera

APPLICATIONS

Photography, environmental scanning

TYPES / SIZES

Vary

ENVIRONMENTAL

May be used for thermal imaging to analyze heat loss

LIMITATIONS

Requires long exposure to record a single image

FUTURE IMPACT

Spatiotemporal explorations of light and environmental effects

COMMERCIAL READINESS

CONTACT

Mediated Matter Group
Massachusetts Institute of Technology
Media Lab
75 Amherst Street, Room E14-433B,
Cambridge, MA 02139-4307
617-324-3626
matter.media.mit.edu
ked03@media.mit.edu

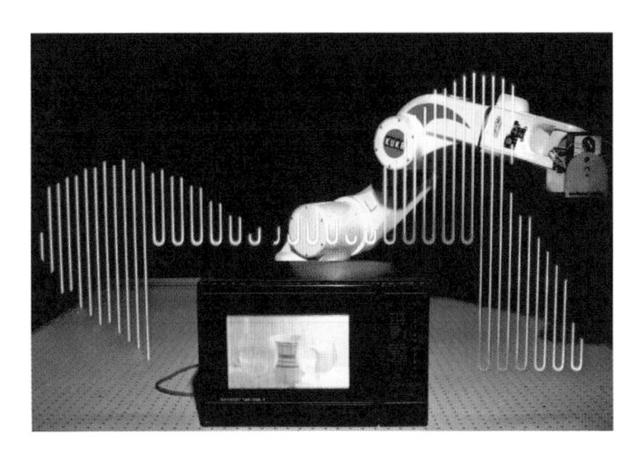

SOLAR DISPLAY INTERFACE

LightByte

LightByte is an interactive optical scrim that modulates solar illumination to create detailed light projections. Created by former MIT Media Lab scholar Sheng-Ying Pao, the solar-based interface transforms sunlight into an expressive medium for conveying information, communicating ideas, and creating intricate patterns.

Hundreds of individually controlled, wood-based servo flaps act like solar pixels, enabling light to be emitted or shaded in varying degrees. LightByte utilizes computing algorithms and behind-the-scenes kinetic mechanisms to animate light based on a variety of user-directed programs. A marriage of a precise digital display and transient daylight, LightByte encourages a provocative dialogue between the synthetic and the natural.

CONTENTS

Wood frame, wood shutters, polymer hinges with metal hardware, kinetic mechanisms, cables, custom software

APPLICATIONS

Active daylight control, digital display, remote communications interface

TYPES / SIZES

35 × 46" (89 × 117 cm) panel with eight kinetic panel modules and 384 individually controlled shutters

ENVIRONMENTAL

Can be programmed for automatic daylight management as an active filter for solar heat gain

LIMITATIONS

Makes audible noise when shutters move

FUTURE IMPACT

Use of sunlight as a vehicle for digital displays, integration of solar control and digital communications technologies

COMMERCIAL READINESS

CONTACT

MIT Media Lab Changing Places
77 Massachusetts Avenue, Room E14/E15,
Cambridge, MA 02139
617-253-5960
cp.media.mit.edu

INTERACTIVE LED PANEL SYSTEM

Mnemosyne

PROCESS

Mnemosyne is a modular interactive LED system for expansive architectural surfaces. Part of Oslo-based Nullohm's Respondent Elements range, Mnemosyne is the latest iteration of the company's Dobpler interfaces. Mnemosyne responds to the movements of passersby in real time, mapping their silhouettes as light shadows.

The robust system is suitable for exterior use in high-traffic areas, particularly for the purpose of improving personal safety and comfort. Nullohm installed the system under a bridge in an Oslo neighborhood with measurable success: the surveilling light map now tracks individuals while illuminating what was once a dark space. As a result, vandalism in the area has declined, and local residents take pride in their own interactive public art installation.

CONTENTS

Printed circuit boards, LEDs, aluminum-PE-and-acrylic composites

APPLICATIONS

Interactive surfaces, street lighting, public art

TYPES / SIZES

7 × 7 × 1/8" (18 × 18 × 0.4 cm) lighting elements in enclosed modules

ENVIRONMENTAL

Low-energy illumination, 100 percent recyclable components

TESTS / EXAMINATIONS

First installation in continuous use since 2006

LIMITATIONS

Direct sunlight will block the sensors

FUTURE IMPACT

Responsive illuminating surfaces used to enhance public safety

COMMERCIAL READINESS

CONTACT

Nullohm AS
Konowsgate 5, Oslo 0192, Norway
+47 92258534
www.nullohm.com
sales@nullohm.no

FLEXIBLE INTERCONNECTED LIGHTING SYSTEM

Non Linear

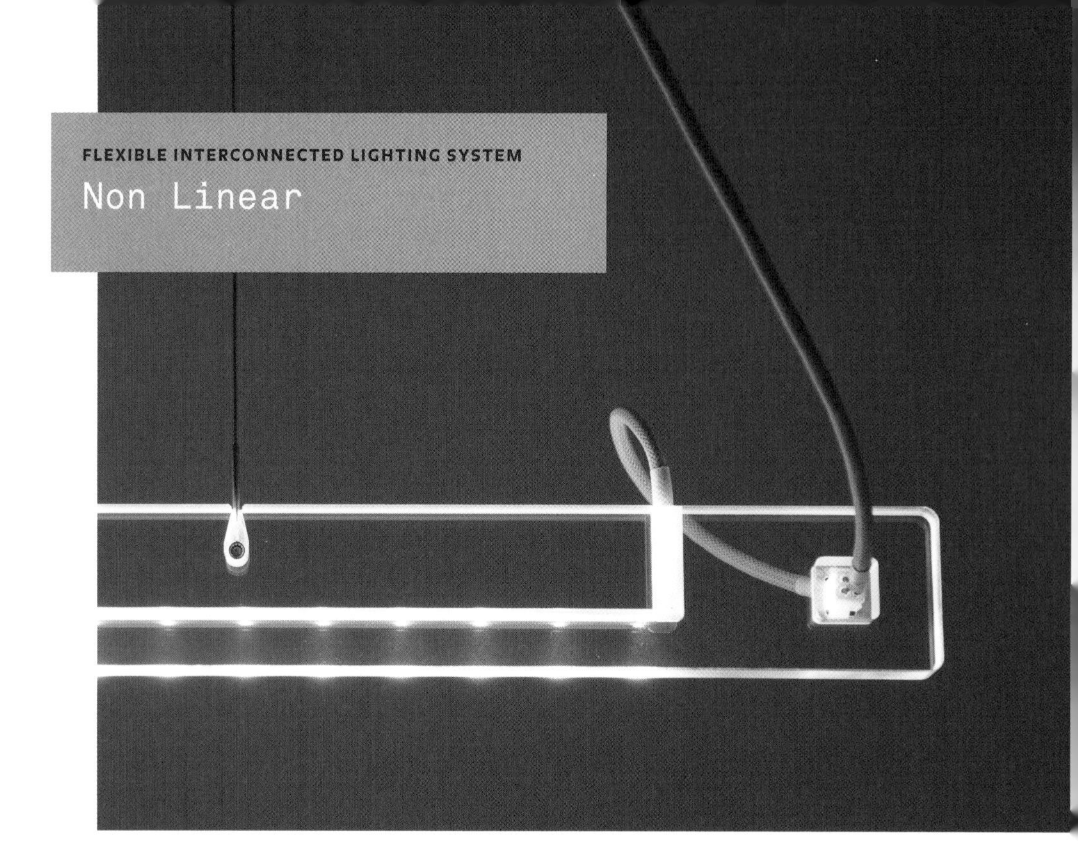

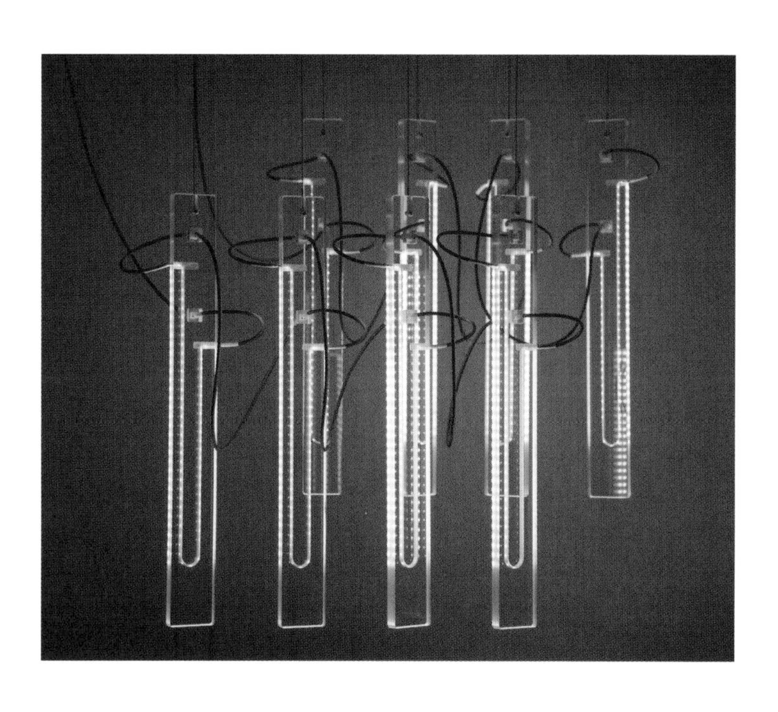

Non Linear is a system of modular LED luminaires made of transparent acrylic. The individual lighting components consist of 1"-thick (2.54 cm) machined material with inlaid LED bands. The collection consists of full-color RGB LEDs that support remote-controlled adjustments via smartphone. Non Linear modules may be connected in various combinations with colorful cords offered in thirty different hues. NONdesigns continues to expand the collection of Non Linear module designs, all of which are now UL listed for commercial and residential applications.

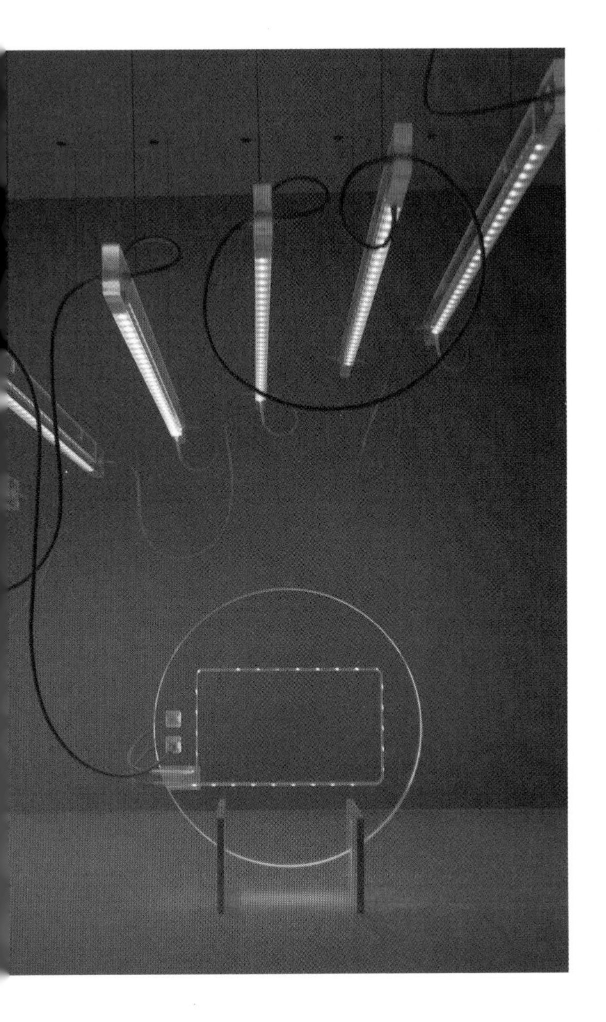

CONTENTS

Acrylic, LED strips, electrical cords

APPLICATIONS

Lighting, illuminated decorative objects

TYPES / SIZES

Adjustable white or RGB color LED models; cords available in 30 color options; H (horizontal) series, V (vertical) series, F (fan) series, T (ovoid) series, W (wall) series; up to 7' 11" (2.4 m) long

ENVIRONMENTAL

Low energy consumption

TESTS / EXAMINATIONS

UL listed

FUTURE IMPACT

Proliferation of DIY, flexible low-power lighting; dematerialization of lighting elements and systems

COMMERCIAL READINESS

●●●●●

CONTACT

NONdesigns LLC
299 North Altadena Drive, Unit 120,
Pasadena, CA 91107
626-696-3081
www.nondesigns.com
info@nondesigns.com

AUTOLUMINESCENT PLANT

Starlight Avatar

Biotechnology company Bioglow has produced the first genetically engineered light-emitting plant. Like glowworms, fireflies, and other naturally occurring autoluminescent organisms, the plant can produce light continuously throughout its life. After the publication of his article "Autoluminescent Plants" in *PLOS ONE* in 2010, molecular biologist Alexander Krichevsky founded Bioglow with the aspiration to create foliage that can double as low-energy light sources.

To make a glowing plant, he introduced the autoluminescent pathway from marine bacteria into the chloroplast genome of the ornamental Nicotiana alata. After engineering a successful if dim specimen, Krichevsky set out to increase the light emission of his plants. In December 2013, Bioglow announced the commercial release of its first decorative autoluminescent plant, the Starlight Avatar, named for its subtle astral radiance.

CONTENTS

Genetically engineered ornamental *Nicotiana alata* plant

APPLICATIONS

Ornamental landscaping, nocturnal wayfinding

TYPES / SIZES

Vary depending on plant species

ENVIRONMENTAL

Energy conservation

TESTS / EXAMINATIONS

USDA

LIMITATIONS

For use in the United States only

FUTURE IMPACT

Genetically modified, multifunctional landscapes; zero-energy autoluminescent lighting

COMMERCIAL READINESS

CONTACT

Bioglow LLC
7568 West Bruno Avenue, St. Louis, MO 63117
631-721-5325
www.bioglowtech.com
info@bioglowtech.com

INTERACTIVE ILLUMINATED PLAYSCAPE

Swing Time

Swing Time is an illuminated playscape consisting of twenty interactive swings. Designed to provide an ad hoc city park by activating unused urban spaces, Swing Time offers three different sizes of light-emitting, ring-shaped swings made from welded polypropylene. Internal microcontrollers adjust the lighting in each swing based on the intensity of its movement, which is measured by an accelerometer. Stationary swings glow with a soft white radiance, while moving swings emit a bright purple light. In this way, Swing Time's interactive elements welcome communities to engage with the playscape and with each other, transforming underutilized sites into laboratories for the stimulation of social activity.

CONTENTS

Welded polypropylene tubes, LEDs, microcontrollers, accelerometer

APPLICATIONS

Playgrounds, public parks, and other spaces; interactive artwork

TYPES / SIZES

Twenty swings, three different sizes

TESTS / EXAMINATIONS

Public installation at the Lawn on D, adjacent to the Boston Convention and Exhibition Center

FUTURE IMPACT

Activation of underutilized public spaces with kinetic and interactive lighting installations, enhancement of public safety by promoting cooperative play

COMMERCIAL READINESS

CONTACT

Höweler and Yoon Architecture LLP
150 Lincoln Street, 3A, Boston, MA 02111
617-517-4101
www.hyarchitecture.com
info@hyarchitecture.com

IMMERSIVE AQUEOUS PROJECTION

Waterlicht

Waterlicht allows viewers to witness the impossible: a flooded Amsterdam—without water. The immersive light painting is a virtual flood, demonstrating what Netherlands would be like without dikes. The artwork consists of undulating lines of light created with LED technology, software, and projection systems that animate low-lying mists.

Conceived as an environmental work of art, Waterlicht was designed to commemorate the Rijksmuseum's 2015 acquisition of the seventeenth-century painting of the Amsterdam flood of 1651 by artist Jan Asselijn. Like the painting, the immersive display reminds the Dutch of their country's sub-sea level position, as well as the universal power of the ocean. It also demonstrates technology's capacity to engulf a large area with light and visual effects: a painting scaled to the city.

CONTENTS

LEDs, projection equipment, AV software

APPLICATIONS

Video projection, landscape lighting, public art

TYPES / SIZES

Covers a 4-acre area

TESTS / EXAMINATIONS

Exhibited in Westervoort and Schokland, Netherlands; Amsterdam; and Paris

LIMITATIONS

Street lighting reduces the visual effect

FUTURE IMPACT

Further development of immersive digital projection and lighting environments at an urban scale

COMMERCIAL READINESS

CONTACT

Studio Roosegaarde
Vierhavensstraat 52–54,
Rotterdam 3029 BG, Netherlands
+31 103070909
www.studioroosegaarde.net
mail@studioroosegaarde.net

Digital

10

> Material Intelligence

> Contemporary society's zeal for
> digital technology might be inter-
> preted as a desire to escape the
> physical world. Our current "informa-
> tion age" and "network society," as
> described by media scholars such
> as Manuel Castells and Frank Webster,
> are defined by an intensified inter-
> est in – and reliance upon – digital
> media. "Indeed, the shift to an
> information society is, often, a
> shift from material objects to digi-
> tal equivalents on computer screens,"
> declare computer scientists Paul
> Dourish and Melissa Mazmanian.[1] In
> *Being Digital*, architect Nicholas
> Negroponte describes this phenomenon
> as a progressive trend: "It is almost
> genetic in its nature, in that each
> generation will become more digital
> than the preceding one."[2]

Yet the digital realm owes its existence to matter. From microprocessors to undersea network cables, virtuality is ever tied to materiality, and there is no escaping this connection. As Dourish and Mazmanian assert, "The information that undergirds the 'information society' is encountered only ever in material form." [3] This awareness is shifting prevalent attitudes toward digitality, with a renewed interest in human-material interactions. The authors declare that "increasingly, social scientists have turned their attention to the intertwining of social phenomena with the material world, arguing both that the social world manifests itself in the configuration and use of physical objects and that the properties of those physical objects and the materials from which they are made—properties like durability, density, bulk, and scarcity—condition the forms of social action that arise around them." [4]

This transformation is popularly known as the *Internet of Things* (IoT), a term coined in 1999 by MIT Auto-ID Center cofounder Kevin Ashton to describe the burgeoning connections between computing and physical objects.[5] Early IoT technologies emphasize the migration of the computer interface to myriad locations. Wearable activity trackers, weight scales, automated household electrical systems, and environmental monitoring networks all incorporate, or otherwise rely upon, screens for the communication of information. Scientist David Rose calls the increasingly prevalent screen "the glass slab," an interface with a predictable future trajectory of more pixels, reduced materiality, and decreased cost.[6] The screen is an eminently useful and familiar medium, but merely embedding it in various places does not exploit the full potential of social-material interactions, or capitalize on humanity's deep evolutionary history connecting with physical objects and natural phenomena. According to Rose, the next technological trajectory will be defined by *enchanted objects*, artifacts that incorporate digital technology in new and surprising ways.[7] This trend may be seen in recently developed interfaces, environments, and fabrication processes that attract our curiosity about the physical world and our desire to engage it in a tactile way. "I simply believe that the most promising and pleasing future is one where technology infuses ordinary things with a bit of magic to create a more satisfying interaction and evoke an emotional response," Rose states.[8]

Researchers at MIT's Tangible Media Group are pursuing this trajectory in what they call *radical atoms,* "a computationally transformable and reconfigurable material that is bidirectionally coupled with an underlying digital model (bits) so that dynamic changes of physical form can be reflected in digital states in real time, and vice versa." [9] **← This concept is evident in dynamic shape displays such as inFORM and Kinetic Blocks, multidimensional interfaces that replicate the dimensionality and movement of remote objects using fields of small actuator-driven blocks.**[10] In 2014 German company Metaio (now owned by Apple) developed the tactile interaction-based technology Thermal

Touch, a process that transforms physical objects into touch screens.[11] The technology employs thermal-imaging cameras to detect the heat signature of users' hands and their proximity to other objects, enabling a variety of touch-actuated processes. Felix Faire's Contact similarly expands the territory of tactility, allowing users to convert any hard surface into a touch-responsive sound and light instrument (see page 294).

Korea-based Jonpasang's Hyper-Matrix interprets the radical atom concept in spatial terms, with a massive three-sided dynamic display that is 148' (45 m) wide and 26' (8 m) tall (see page 302). Luftwerk's Flow likewise spatializes information using programmed light projection with mist screens in urban environments (see page 300). The responsive installation transforms user interaction with light into a spatial, immersive experience. **↓ The Aerial 3D Display developed by Burton employs plasma emission and lasers to create volumetric projections of light dots in space (see page 288).** The company's Super Real Vision display is a compact version that also spatializes virtual models and animations using laser optics in combination with a chiller.[12]

Next-generation material fabrication and automation are also territories of "enchantment." For example, **→ 4D Printing is a process that utilizes multimaterial, three-dimensional prints that automatically change shape over time (see page 286). Composed of both rigid and dynamic parts, 4D Printing assemblies morph from their original printed shapes into geometric structures when activated by water or other energy sources.** M-Blocks are intelligent cubes that work together to assemble various constructions (see page 304). The colorful blocks at first appear to be simple, inert objects, as they have no external moving components. Yet internally rotating flywheels employ angular momentum to spin the cubes while discreet anchoring magnets enable them to align precisely with their neighbors. Larger structures are made possible by flying quadrotor helicopters that construct

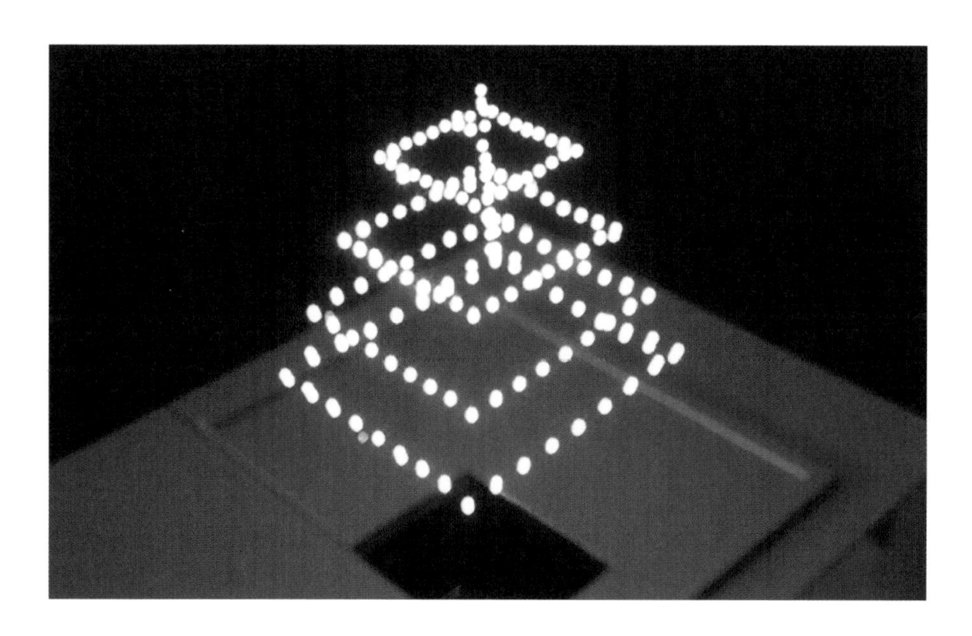

complex assemblies of foam blocks based on preprogrammed algorithms (see page 298).

The concept of enchanted objects represents the full potential of the Internet of Things: the seamless fusion of atoms and bits. In *Understanding Media: The Extensions of Man*, Marshall McLuhan presaged electric technology being an extension of the human nervous system: "With the arrival of electric technology, man extended, or set outside himself, a live model of the central nervous system itself." [13] The notion that this system may be conjoined with the physical world of designed objects promises a new scale of human agency, along with increased complexity. In *Shaping Things*, author Bruce Sterling proffers the term *spime* to describe an object that may be monitored in space and time throughout its life cycle (much like humans are today).[14] Spimes are the tools of what Sterling calls a *synchronic society*, a population that exercises full authority over material histories: "In a synchronic society, every object worthy of human or machine consideration generates a small history. These histories are not dusty archives locked away on ink and paper. They are informational resources, manipulable in real time." [15] Like spimes, enchanted objects will exhibit increasingly lifelike properties, and these will collectively form enchanted systems—higher-level techno-material ecosystems in which all designed objects have enchanted properties. According to Rose, this "New Ecosystem" will have "the potential to take us into terra incognita, unknown and uncharted technological and human territory." [16]

---

1 Paul Dourish and Melissa Mazmanian, "Media as Material: Information Representations as Material Foundations for Organizational Practice" (working paper for the Third International Symposium on Process Organization Studies, Corfu, Greece, June 2011), 2–3.
2 Nicholas Negroponte, *Being Digital* (New York: Knopf, 1995), 231.
3 Dourish and Mazmanian, "Media as Material," 3.
4 Ibid.
5 Kevin Ashton, "That 'Internet of Things' Thing," *RFID Journal*, June 22, 2009, http://www.rfidjournal.com/articles/view?4986.
6 David Rose, *Enchanted Objects: Innovation, Design, and the Future of Technology* (New York: Simon & Schuster, 2015), 11.
7 Ibid., 13.
8 Ibid.
9 See accessed January 21, 2016, http://tangible.media.mit.edu/vision/.
10 See accessed January 21, 2016, http://tangible.media.mit.edu.
11 See accessed January 21, 2016, https://www.metaio.com.
12 See accessed January 21, 2016, http://www.burton-jp.com/en/product.htm.
13 Marshall McLuhan, *Understanding Media: The Extensions of Man* (Cambridge, MA: MIT Press, 1964), 43.
14 Bruce Sterling, *Shaping Things* (Cambridge, MA: MIT Press, 2005), 43.
15 Ibid., 45.
16 Rose, *Enchanted Objects*, 263, 210.

MULTIMATERIAL SHAPE CHANGE

4D Printing

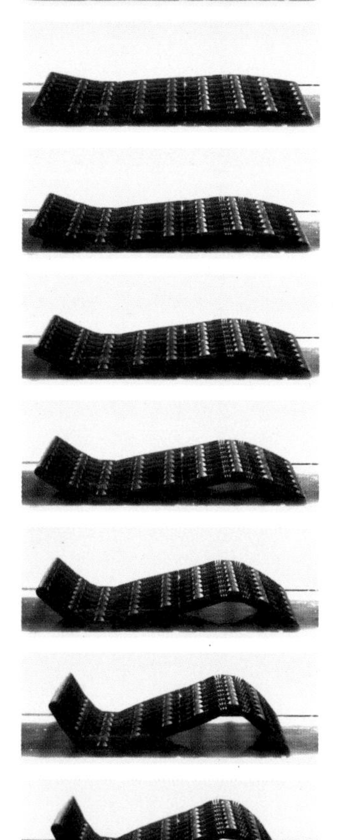

4D Printing is a method for printing customizable smart materials that transform over time. Developed through a collaboration between MIT's Self-Assembly Lab, Stratasys, and Autodesk, 4D Printing employs the Stratasys Connex printer to create multimaterial objects that can shape-shift immediately after printing. This transformative process enables enhanced functionality in materials, including sensing, actuation, and assembly. Possible uses include robotic approaches without the requirement for mechanical systems, in addition to responsive apparel, objects, or surfaces that can shift according to user-based or environmental needs.

The Self-Assembly Lab creates multimaterial prints that consist of single strands or flat plates that morph into sophisticated three-dimensional objects when activated via water, light, heat, or another energy source. The key to 4D Printing is geometry. A hydrophilic polymer that is activated by water can provide the necessary expansion capabilities, for example, but the resulting form must first be designed with knowledge of the material's behavior.

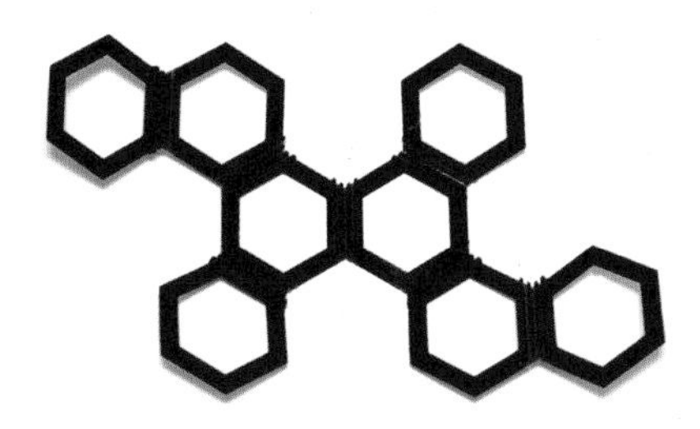

CONTENTS

Hydrophilic polymer or another active component, rigid polymer or another inert material

APPLICATIONS

Self-assembling structures, responsive garments, adaptive products

TYPES / SIZES

Hydrophilic material expands to 150 percent of its original size in water; prototype strands of 12" (30 cm), 18" (46 cm) length; Hilbert curves of 14' (4.3 m), 50' (15 m) length

ENVIRONMENTAL

Low energy construction

LIMITATIONS

Requires an energy input for activation

FUTURE IMPACT

Automated assembly of buildings and products from multimaterial surfaces; improved worker safety, reduced construction errors

COMMERCIAL READINESS

CONTACT

Massachusetts Institute of Technology
School of Architecture + Planning
Self-Assembly Lab
77 Massachusetts Avenue, Cambridge, MA 02139
617-253-7791
www.selfassemblylab.net
info@selfassemblylab.net

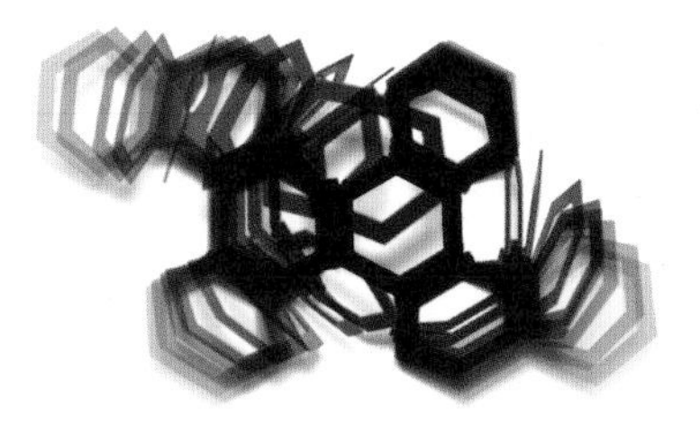

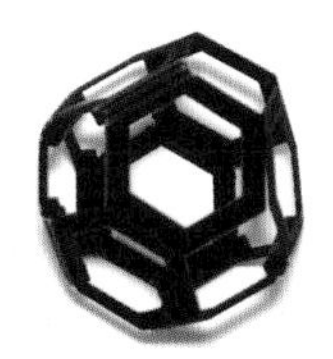

SPATIAL DIGITAL PROJECTION SYSTEM

Aerial 3D

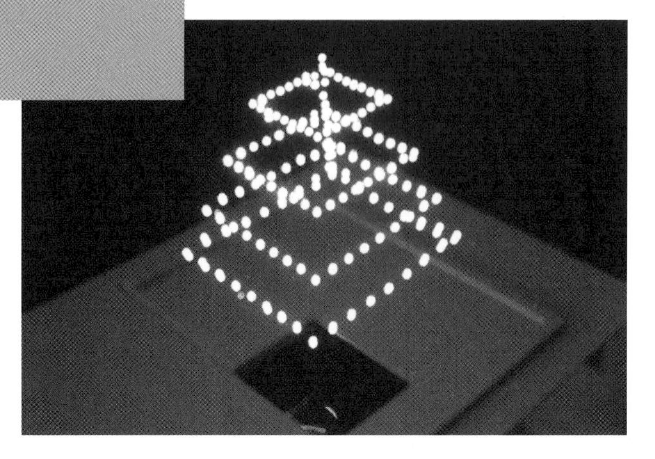

PROCESS

Most three-dimensional displays create pseudo-3D images on 2D planes by taking advantage of a perceptual phenomenon called human binocular disparity. However, there are many challenges to this method, including the limitation of the visual field and the lack of any true spatial representation of three-dimensional information.

To overcome these limits, Japanese company Burton has developed a real 3D display composed of bright dots of light that appear in midair. The display device employs a plasma emission process using focused laser light. Precise control over the laser's spatial targeting enables the projection of a collection of light points, creating three-dimensional images out of dot arrays constructed in the air. In this way, users can perceive 3D imagery in true three-dimensional space.

CONTENTS
Laser, optics, chiller

APPLICATIONS
Video projection, digital signage, emergency communication

TYPES / SIZES
Scalable to space

ENVIRONMENTAL
Minimal use of resources: no projection screen required

LIMITATIONS
Monochromatic, produces audible static noise

FUTURE IMPACT
Capacity of digital information to become spatial, enhancement of emergency communications

COMMERCIAL READINESS

CONTACT
Burton Inc.
1–2–10 Hamacho, Kawasaki-ku, Kawasaki 210-0851, Japan
www.burton-jp.com
info@burton-jp.com

FLUID-FILLED TACTILE SENSOR

BioTac

PROCESS

BioTac emulates the sensory capability and physical qualities of the human fingertip. The design consists of a rigid core enveloped by a liquid-filled, elastic skin with sophisticated tactile sensing. Developed by SynTouch to provide robots with haptic awareness, BioTac enables machine touch via sensing force, vibration, and temperature.

Standard tribology technology, which pertains to the interaction between moving surfaces, produces calculable measurements; however, these are not typically correlated with human tactility. Designers must, therefore, depend upon the subjective opinions of consultants and focus groups for information about touch. Using BioTac as the foundation of its testing system, SynTouch employs a process that combines precise quantification with humanlike perception. The company views this technology as a critical part of connecting human and machine awareness.

CONTENTS
85 percent epoxy resin, 7 percent silicone elastic rubber, 5 percent PCB and integrated circuits, 2 percent polyethylene glycol, 1 percent platinum

APPLICATIONS
Robotics, prosthetics, haptic sensing

TYPES / SIZES
$2 \times 1/2 \times 1/2$" (51 × 13 × 13 mm)

LIMITATIONS
Maximum bending force of 50 N

FUTURE IMPACT
Endowing robotics systems with increased lifelike capacities, advancing the science of somatosensory systems and prosthetic technologies

COMMERCIAL READINESS

CONTACT
SynTouch LLC
2222 South Figueroa Street PH2,
Los Angeles, CA 90007
213-493-4400
www.syntouchllc.com
info@syntouchllc.com

FREEFORM 3D WALL PRINTING

C-Fab

Branch Technology has developed a systemic approach to 3D printing mesh structures that function as the backbone of a wall assembly. The Cellular Fabrication (C-Fab) process is based on the knowledge that composite structures, such as human bones and trees, are common in nature. In the C-Fab structures, optimized geometries are imparted strength and functionality by the mesh material, allowing exterior and interior surfaces to be finished in any fashion.

C-Fab is made possible by a free-form printing process that is not constrained by the slow, layer-upon-layer manufacturing method of traditional 3D printing. The printing head is connected to a 12.5' (3.8 m) robotic armature that slides along a 33' (10.1 m) rail and is capable of printing walls up to 58' (17.7 m) in length. Branch Technology's algorithm generates both the spatial data and robot-driven motion to build complex geometric structures without the need for support scaffolds. The free-form nature of the approach encourages architects and designers to explore dynamic forms as well as their structural implications, while minimizing the amount of material employed.

CONTENTS

ABS plastic, carbon fiber

APPLICATIONS

Construction of lightweight shelters, walls, furniture, pavilions, and sculptural objects

TYPES / SIZES

Can print a 3D matrix up to 25' (7.6 m) wide and 58' (17.7 m) long

ENVIRONMENTAL

Cellular construction optimizes material resources, and strength-to-weight ratio

LIMITATIONS

Currently focused on wall construction, not yet used to fabricate roofs

FUTURE IMPACT

3D printing of composite structures for whole buildings

COMMERCIAL READINESS

CONTACT

Branch Technology
100 Cherokee Boulevard, Suite 125,
Chattanooga, TN 37405
333-224-9495
www.branch.technology

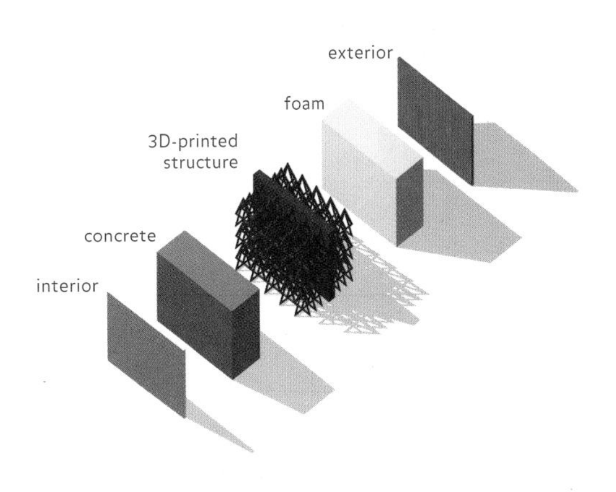

INTERACTIVE SURFACE WITH AUGMENTED ACOUSTICS

Contact

Contact transforms any hard surface into a sound and light instrument. Developed by Felix Faire while a student in Bartlett's Interactive Architecture Lab, the interface employs contact piezoelectric microphones and waveform analysis to enable a tabletop or other surface to be used as a touch-sensitive device.

Using predetermined algorithms, Contact translates finger and hand taps into audio impulses, and a custom loop pedal enables sound repetitions. Arduino and piezo sensors are also employed to translate tactile pressure into audio volume. Visualizations of the resulting waveforms are projected onto the interactive surface, creating an immersive audiovisual experience.

CONTENTS

Contact microphone, computer, solid surface

APPLICATIONS

Interactive surfaces, environmental instruments

TYPES / SIZES

Vary with context and application

ENVIRONMENTAL

Enables the interactive use of existing objects without the need to embed complicated electronics into new products

LIMITATIONS

The system must be trained for each application, the accuracy of interaction recognition varies with different materials

FUTURE IMPACT

Immersive, tactile interfaces; augmented interactivity based on reuse of existing inert objects

COMMERCIAL READINESS

● ○ ○ ○ ○

CONTACT

Felix Faire
London N17HL, UK
+44 07584096468
www.felixfaire.com
felixfaire@gmail.com

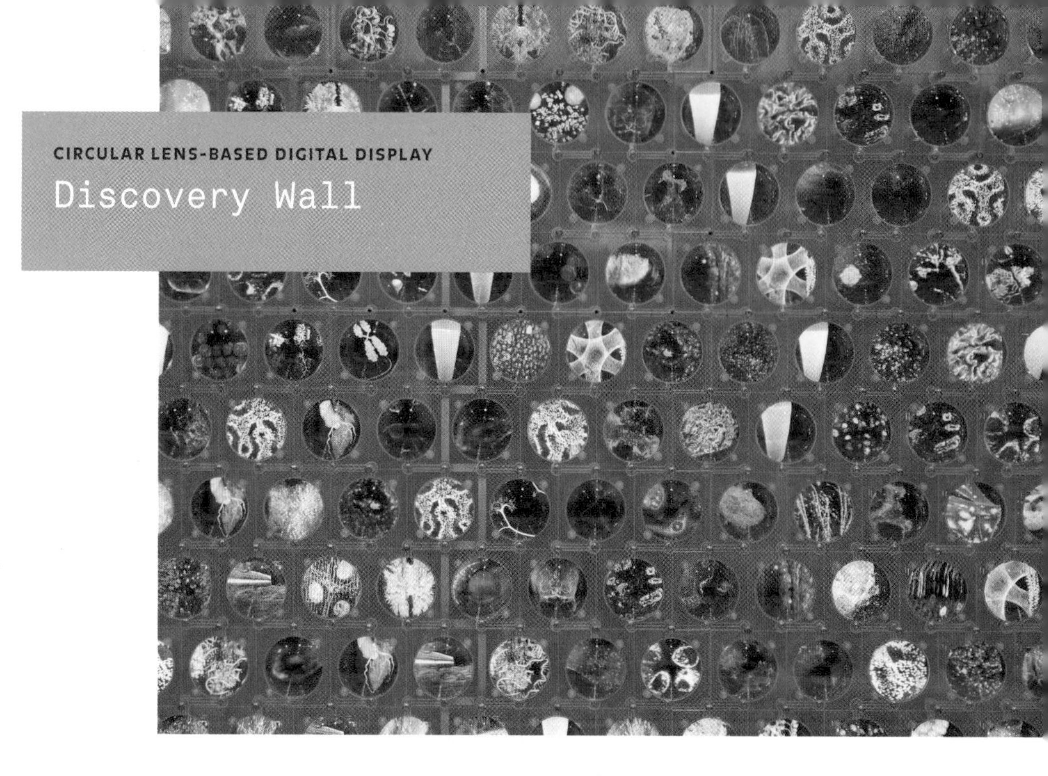

CIRCULAR LENS-BASED DIGITAL DISPLAY

Discovery Wall

The Discovery Wall is a display wall composed of circular pixels. Mini LCD screens are set in a grid pattern behind circular Fresnel lenses, which magnify the screen images. The screens broadcast continuously changing images and text via content management system (CMS) software. The screens can display content individually at a micro level or collectively at a macro level. The system is therefore easily adaptable to a range of audience speeds and distances, from slow-moving pedestrian window shoppers to passing drivers.

The Discovery Wall can be installed as part of an interior facade at various scales and sizes. The screens are attached to a printed circuit board that can be customized according to the building environment. Live information can be sent to the screens from a range of platforms, and if required, the CMS can be connected to APIs that allow users to engage with the wall and upload their own content.

CONTENTS

LCD screens (240 × 240 pixel resolution), Fresnel lenses, printed circuit board, attachments, CMS software

APPLICATIONS

Display wall, educational interface, feature wall

TYPES / SIZES

The circuit board size and color can be customized to fit a range of applications, content templates can be created according to user requirements

ENVIRONMENTAL

2,800 LCD screens collectively use less than 1 kW of power, the screens are manufactured from recycled LCD displays

TESTS / EXAMINATIONS

2014 operational installation at the Belfer Research Building in New York

LIMITATIONS

Not suitable for unprotected outdoor use (requires glass facade)

FUTURE IMPACT

Digital display surfaces are expected to proliferate widely, the Discovery Wall represents a refreshingly physical and multidimensional display platform

COMMERCIAL READINESS

CONTACT

Squint / Opera with Hirsch & Mann
1–5 Vyner Street, London E2 9DG, UK
+44 02079787788
www.squintopera.com
hello@squintopera.com

FLYING MACHINE–ENABLED CONSTRUCTION

Flight Assembled Architecture

Flight Assembled Architecture is a process in which flying machines construct prototype building structures. Developed by ETH Zürich professor Raffaello D'Andrea and architecture firm Gramazio Kohler Architects, Flight Assembled Architecture employs software-controlled flying robots to place individual foam bricks in order to construct a large open-weave structure.

Imagined as a scale representation of a 1,969'-tall (600-m) towering city for 30,000 inhabitants, the installation consists of 1,500 prefabricated polystyrene foam modules and stands 19.7' (6 m) high and 11.5' (3.5 m) in diameter at 1:100 scale. The quadrotor helicopters required several days of continuous flight to build the megastructure simulation. The potential to use flying machines for building construction has many compelling benefits, such as ensuring human safety and precision of craft.

CONTENTS

Quadrotor helicopters, polystyrene foam building modules

APPLICATIONS

Automated construction

TYPES / SIZES

Initial assembly made of 1,500 modules, 19' 8" (6 m) high × 11' 6" (3.5 m) diameter

TESTS / EXAMINATIONS

Working prototype installed at the FRAC Centre in Orléans, France

LIMITATIONS

Weight poses a challenge to air-based building construction, so light building materials are a priority

FUTURE IMPACT

Digitally automated, drone-based building construction would increase worker safety and enhance the ability to construct complex geometries

COMMERCIAL READINESS

● ○ ○ ○ ○

CONTACT

ETH Zürich
Institute for Dynamic Systems and Control
Leonhardstrasse 21, LEE K, Zurich 8092, Switzerland
+41 446326596
www.idsc.ethz.ch
mavt@mavt.ethz.ch

ILLUMINATED RESPONSIVE WATER SCREENS

Flow

Flow is an art installation that uses one or more custom-designed aluminum square tubes with embedded brass spray nozzles that generate a fine mist of water. The constant flow of water creates a veil or curtain effect, providing a screen on which to project imagery.

Originally conceived by Luftwerk to celebrate the twentieth anniversary of the sister-city relationship between Chicago and Hamburg, Germany, *Flow* is inspired by the element of water and its all-encompassing connectivity. The designers first revealed the approach in the outdoor public art exhibition *FLOW / Im Fluss*, a visualization of the characteristics of the cities' two rivers: Hamburg's Elbe and the Chicago River. Luftwerk projected information concerning river health—including chemical pollution and dissolved oxygen levels—onto the mist screens, creating an immersive, three-dimensional display.

CONTENTS
Water, sensors, custom-built square tubing with welded spray nozzles, light sources

APPLICATIONS
Interactive art, exterior water feature

TYPES / SIZES
Vary

LIMITATIONS
Interior use requires the provision of drainage

FUTURE IMPACT
Art engaging the public realm; proliferation of interactive, immersive water features

COMMERCIAL READINESS
●○○○○

CONTACT
Luftwerk
860 North California Avenue, Chicago, IL 60622
www.luftwerk.net
info@luftwerk.net

KINETIC PROJECTION SURFACE

Hyper-Matrix

Hyper-Matrix is a kinetic landscape composed of walls of moving blocks. Created for the Hyundai Motor Group Exhibition Pavilion at the 2012 World Expo in Yeosu, South Korea, Hyper-Matrix consists of thousands of 12 5/8 × 12 5/8" (320 × 320 mm) white foam cubes connected to stepper motors within a large steel frame. Actuators adjust the relative depth of each cube, creating patterns across the three-sided display.

Composed of what at first appear to be three blank white walls, the Hyper-Matrix installation quickly comes to life as thousands of individual modules begin to move, pulsate, and form dynamic images across the room. The modules are densely packed, with only 3/16"-wide (5-mm) interstitial spaces, so the surfaces also function well as projection screens. Digital content projected on the walls coordinates with the cubes' movement.

CONTENTS

Foam cubes, actuators, stepper motors, steel frame, controller

APPLICATIONS

Projection screen, information display, interactive surface

TYPES / SIZES

Cube size 12 5/8 × 12 5/8 × 12 5/8" (320 × 320 × 320 mm), 3/16" (5 mm) intervals

TESTS / EXAMINATIONS

Installed at the 2012 World Expo, Yeosu, South Korea

LIMITATIONS

Not designed for exterior use

FUTURE IMPACT

The line between media and architecture will dissolve, and information displays will have an increased physical presence

COMMERCIAL READINESS

CONTACT

Jonpasang
396-37, Mangwon-dong, Mapo-gu, Seoul, South Korea
+82 23327337
www.jonpasang.com
info@jonpasang.com

SELF-ASSEMBLING ROBOTIC CUBES

M-Blocks

PRODUCT

M-Blocks are simple cubic robots that propel themselves through space without any conspicuous moving parts. Developed by scientists at MIT's Computer Science and Artificial Intelligence Laboratory, M-Blocks can push, spin, or roll across a surface. They can climb one another to make three-dimensional structures and even jump from one position to another.

M-Blocks' seemingly magical capabilities are enabled by cleverly hidden mechanics. Each cube includes an internal flywheel that creates angular momentum when it brakes, as well as a battery for power and a radio for communication. The modules use a combination of surface magnets and edge magnets to join precisely with their neighbors, and can form elaborate lattice-like structures collectively.

Although commands are now sent remotely, the researchers' goal is to impart the modules themselves with mobility intelligence so that they can make autonomous decisions based on environmental conditions. The animate cubes could repair structures, self-construct scaffolding, or mobilize within zones that are unsafe for humans.

CONTENTS

Beveled aluminum frame, flywheel, brushless motor controller, braking mechanism, electronics, radio, battery, face magnets, edge magnets, face plates

APPLICATIONS

Structural repair, self-configuring scaffolding, self-assembling furniture, environmental monitoring, building in hazardous conditions

TYPES / SIZES

1 15/16 × 1 15/16 × 1 15/16" (50 × 50 × 50 mm)

TESTS / EXAMINATIONS

Motion, endurance, material surface tests

LIMITATIONS

Modules only contain a single, unidirectional actuator, and they are not yet truly autonomous, since they must receive algorithm commands via radio

FUTURE IMPACT

Mobile self-assembling robots will extend construction capabilities beyond the factory floor, enabling structural repair and automated construction—particularly in challenging conditions

COMMERCIAL READINESS

● ○ ○ ○ ○

CONTACT

MIT Computer Science and Artificial Intelligence Laboratory
32 Vassar Street, Cambridge, MA 02139
617-253-5851
www.csail.mit.edu
webmaster@csail.mit.edu

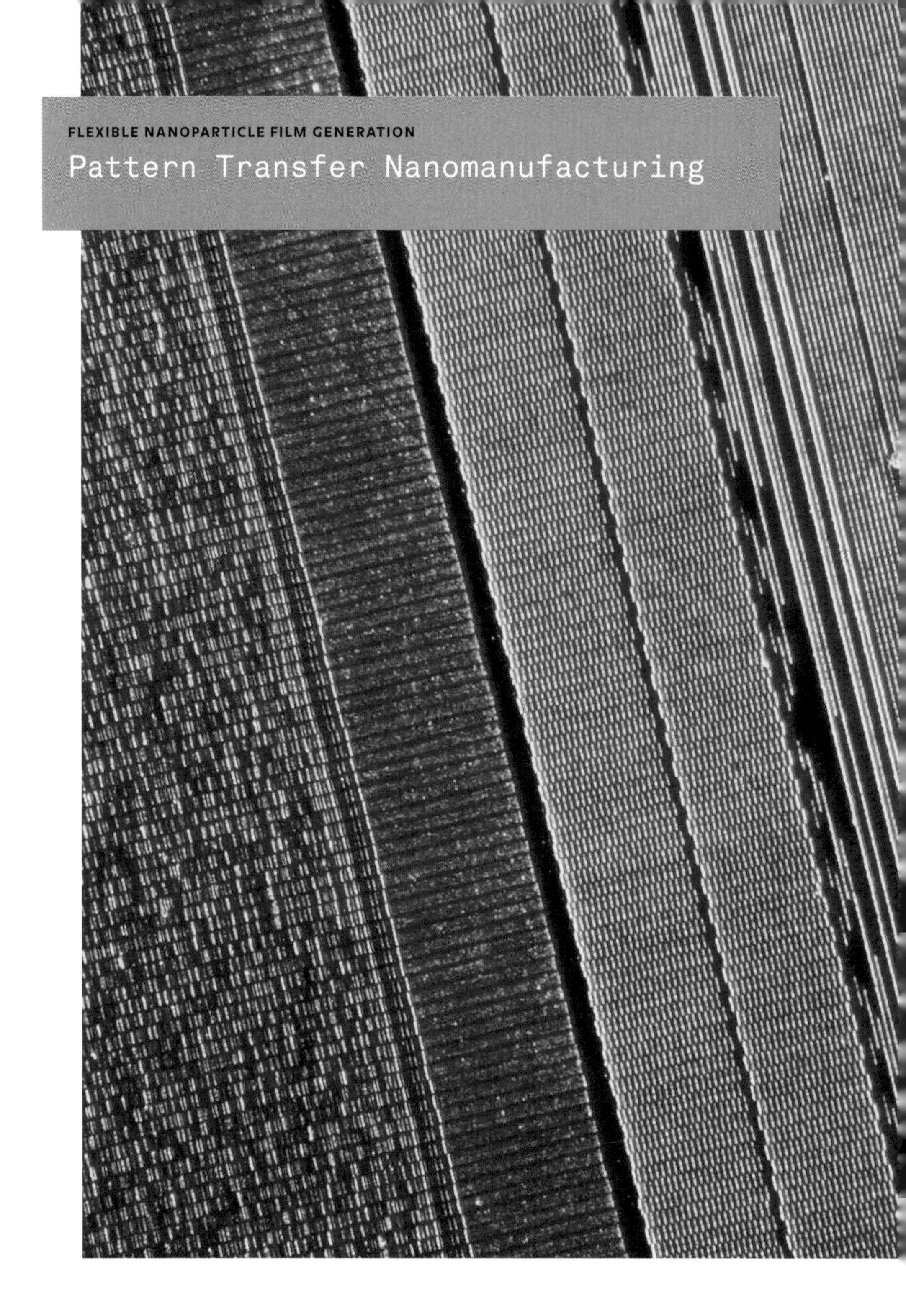

FLEXIBLE NANOPARTICLE FILM GENERATION

Pattern Transfer Nanomanufacturing

PROCESS

University of South Carolina scientists have developed an innovative "misuse" of magnetic recording technology with intriguing possibilities. The technology is conventionally used for data storage; the new application fabricates sophisticated materials by modifying individual nanoparticles. Employing a technique called Pattern Transfer Nanomanufacturing, physicist Thomas Crawford and his team have transformed the "disk" from a disk drive into a reprogrammable template. They first suspend magnetic nanoparticles in a liquid known as a ferrofluid. Once they coat the disk with ferrofluid, they pull the nanoparticles to the surface using a magnetic field. When the nanoparticles have assembled into place, the team removes the fluid and finishes the disk surface with a liquid polymer. They then peel the cured solid polymer from the disk to yield a flexible and transparent film that contains the patterned nanoparticles.

Crawford expects these films will find application in optics, biotechnology, and novel material surfaces that can be programmed for a variety of uses. Examples include manipulable surface films that enable selective light transmission and conductivity, thereby enhancing solar control or communications networks.

CONTENTS

Air-curable liquid polymer

APPLICATIONS

Programmable surfaces, optical and electronic devices

TYPES / SIZES

Vary, approximately 10 nm resolution

TESTS / EXAMINATIONS

Laboratory tests

FUTURE IMPACT

Manufacture of sophisticated polymer films with embedded intelligence

COMMERCIAL READINESS

CONTACT

University of South Carolina
Department of Physics and Astronomy
712 Main Street, Columbia, SC 29208
803-777-8105
www.physics.sc.edu
crawford@physics.sc.edu

PLUG-AND-PLAY BUILDING MONITORING SYSTEM

Pointelist

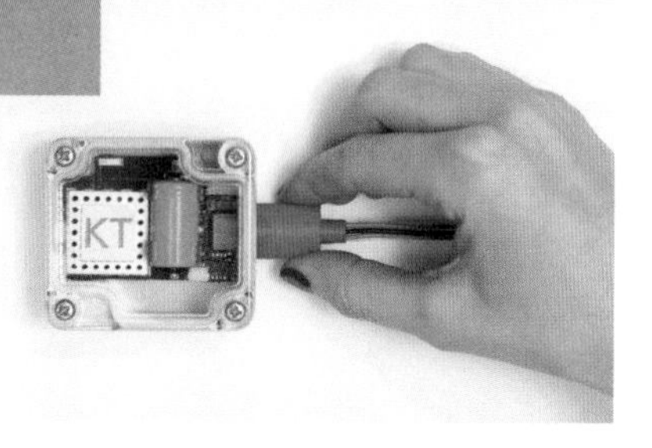

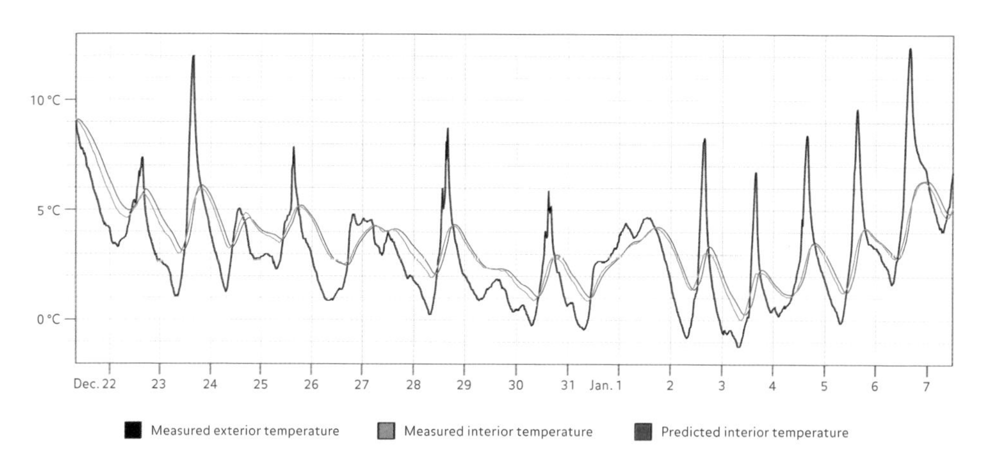

PRODUCT

Researchers at KieranTimberlake have developed a Wireless Sensor Network to monitor building performance and behavior. Seeking to make buildings as climate sensitive and responsive as our bodies, the researchers have created a network of easy-to-install, plug-and-play sensors that can receive data wirelessly for viewing in a web interface, in addition to software tools for analyzing the data. The first iterations of the sensor network focus on thermal characteristics, although other climatological qualities may also be tracked. The goal of the system is not just to produce raw data, but also to help users filter and evaluate the data in the most useful and relevant ways for managing operations over a building's lifetime.

CONTENTS

Temperature sensors, displacement connectors, wireless node, drop cable, pull tab, weathertight fitting

APPLICATIONS

Localized climate monitoring in new and existing buildings

TYPES / SIZES

Wireless node roughly 2 × 2" (5 × 5 cm), component sizes vary

ENVIRONMENTAL

Enables real-time monitoring of climate changes and informed architectural responses

TESTS / EXAMINATIONS

Field tests in Philadelphia, Boston, New Haven, and Minneapolis

LIMITATIONS

Currently requires wired electrical source (battery version under development)

FUTURE IMPACT

Ability for buildings to monitor and communicate specific environmental changes, enabling more reliable design and management decisions

COMMERCIAL READINESS

CONTACT

KieranTimberlake
841 North American Street, Philadelphia, PA 19123
215-922-6600
www.kierantimberlake.com
services@kierantimberlake.com

Material Index

Creator Index

Manufacturer Index

Acknowledgments It's been over a decade since the first *Transmaterial* volume was published, and I am very grateful for the rewarding journey it inspired. After three books in the series, I decided to focus on material applications, specifically, in subsequent research. However, the topic of the next generation of transformative materials has continued to compel me, and I am therefore deeply appreciative for the chance to breathe new life into the *Transmaterial* project. This opportunity would likely not have arisen without the strong support for the books and website, for which I am very thankful.

The University of Minnesota, where I teach, has supported my work in significant ways. I am indebted to my graduate research assistant, Holly Engle, who worked tirelessly to deliver many of the contacts, agreements, and image files necessary for the manuscript submission. I am also thankful to Patrick Becker, an undergraduate research assistant, for his help in this effort. I would like to thank the University of Minnesota School of Architecture and Undergraduate Research Opportunities Program for their financial support of Holly and Patrick, respectively. I would also like to thank architecture faculty Marc Swackhamer, Renée Cheng, and Mary Guzowski for their continued mentorship and support of my research.

I am indebted to the folks at Hanley Wood and *Architect* magazine for the opportunity to write the Mind & Matter column. This regular column keeps me focused on the latest material developments and has therefore helped me to generate the content for this book. I am grateful to my editor, Hallie Busta, as well as Wanda Lau and Ned Cramer, for their continued support.

Transmaterial Next came to life during conversations with Jennifer Lippert, my longtime collaborator and friend at Princeton Architectural Press, just as the first *Transmaterial* volume did in 2005. I would like to express my appreciation for her unwavering commitment to this work. I would also like to thank editorial assistant Simone Kaplan-Senchak, copy editor Lisa Delgado, and editor Jenny Florence—as well as book designer Ben English—for their significant contributions to the endeavor.

Finally, I am deeply grateful to my family—Heather, Blaine, and Davis Brownell—for their unflagging support and encouragement of my work.

Image credits **Front cover:** clockwise from top left: ICD University of Stuttgart; François Lauginie; the Self-Assembly Lab, MIT; Kieran Timberlake; Institute for Advanced Architecture of Catalonia **Back cover:** clockwise from top right: Carmen Trudell & Natacha Schnider; Rosalie McMillan; Jacob Snavely **8:** Steven Keating **10:** top Fraunhofer WKI; bottom University of Stuttgart, Max Planck Institute **11:** Sonice Development **12:** StoneCycling BV **13:** AnaElise Beckman, Alexandra Cohn, and Michael Zaiken **15:** D. Serkentzis **23:** Taj Easton **24:** Rachel Boyd **25:** Loom P.A. **26–27:** Kazushi Miyamoto **28:** top Rachel Boyd; bottom S. Woodhouse **29:** Rachel Boyd **30:** The ACE-MRL, University of Michigan **32:** Escofet 1886 S.A. **34–35:** Thorsten Klooster, Heike Klussmann **36:** top Carmen Trudell & Natacha Schnider; bottom Carmen Trudell **37:** Carmen Trudell & Natacha Schnider **38–29:** Thorsten Klooster, Heike Klussmann **40–41:** Steven Keating **42:** Fraunhofer IBP **44:** University at Buffalo **46:** Jacob Snavely **47:** Ed Caldwell **51:** U.S. Naval Research Laboratory/Jamie Hartman **52:** WASProject **53:** Ordinary Ltd/Magnus Larsson & Alex Kaiser **54:** Zhejiang University **56:** bioMASON **58:** University of Stuttgart, Max Planck Institute **60–61:** Shiro Studio **62:** MIT Mediated Matter **63:** Steven Keating **64–65:** Erik Schlangen, TU Delft **66:** top and bottom left Pim Hendriksen; bottom right Studio Roosegaarde **67:** Studio Roosegaarde **68–69:** Institute for Advanced Architecture of Catalonia **70–71:** U.S. Naval Research Laboratory/Jamie Hartman **72–73:** StoneCycling BV **77:** Dan Little/HRL Laboratories **78:** top Blaine Brownell; bottom MX3D **79:** Lotte Stekelenburg **80–81:** Andrew O. Payne, LIFT architects **82–83:** Höweler and Yoon Architecture, LLP **84–85:** Jinil Park **86–87:** DOSU Studio Architecture **88–89:** University of Utah **90–91:** Kyouei Design **92:** Lotte Stekelenburg **93:** top Sylvain Lefevre; bottom Studio Roosegaarde **94:** Dan Little/HRL Laboratories **96:** top and bottom MX3D; middle Robert Roozenbeek **97:** MX3D **98–99:** Höweler and Yoon Architecture, LLP **103:** Raw Color **104:** top Carolien Laro; bottom Alan Rudie, Forest Service **105:** MIT Mediated Matter **106–107:** D. Serkentzis **108–109:** ICD University of Stuttgart **110:** Canek Fuentes-Hernandez, Georgia Tech **111:** Alan Rudie, Forest Service **112:** Fraunhofer ICT **114–115:** Raw Color **116–117:** MIT Mediated Matter **118:** top Xia Zhi; bottom left and right Penda Architecture **119:** Penda Architecture **120:** top Martin Dee; bottom Ata Sina **121:** Ata Sina **122–123:** Bezalel **124–125:** Carolien Laro **126–127:** Alcarol **128:** top Sandy Wang; bottom Matthew Soules Architecture **129:** Cameron Koroluk **130:** Fraunhofer WKI **132–133:** Jannis Hülsen, Stefan Schwabe **137:** Javier G. Fernandez **138:** NASA Langley Research Center **139:** Waël Seaiby **140:** Martin Jansen **141:** Lilian van Daal **142–143:** Javier G. Fernandez **144:** Rosalie McMillan **145:** Lynton Pepper **146–147:** Steven Keating **148–149:** Andrea Valentini **150–151:** Wiktoria Szawiel **152:** Jeff Fitlow/Rice University **153:** Rafael Verduzco/Rice University **154:** Danielle Storm Hoogland **155:** Yeadon Space Agency **156–157:** Waël Seaiby **158–159:** RVTR **160:** Roberto Martín **162:** top Ryan Lodermeier; bottom HouMinn Practice **164–165:** Zeoform **169:** Andrew Thomas Ryan **170:** top Francois Barthelat/McGill University; bottom Solar Roadways **171:** Steven Keating **172:** Höweler and Yoon Architecture, LLP **174:** Francois Barthelat/McGill University **176:** top Klaus Kerth/Foamglas; bottom Wolfgang Bicheler/Foamglas **177:** Foamglas **178:** top row MIT Mediated Matter; bottom row Andrew Thomas Ryan **179:** MIT Mediated Matter **180:** Fraunhofer ISC **182–183:** Michael Fent **184–185:** Solar Roadways **186–187:** Colt/Arup/SSC **188:** top Patrick H. Corkery; bottom SolarWindow Technologies, Inc. **189:** Ellen Jaskol **190:** AIST **192–193:** Steven Keating **197:** Yeadon Space Agency **198:** top Benjah-bmm27; bottom Sonice Development **199:** Kevin Trung Ha **200:** top left and right Studio Blond & Bieber; bottom left and right Lukas Olfe **201:** Lukas Olfe **202:** Fraunhofer UMSICHT **204–205:** Sonice Development **206–207:** E Ink Holdings Inc. **208–209:** Sonice Development **210:** The Unseen **212–213:** Yeadon Space Agency **214–215:** Studio Marlène Huissoud **216:** Graeme Fleming/University of Strathclyde **218:** Wyss Institute **220:** Pim Hendriksen **225:** The Living **226:** top CSIRO; bottom Axel Körner, Larissa Born **227:** yet|matilde s.n.c. **228:** bottom left Zuzanna Weiss; all others Guillaume Couche **229:** Guillaume Couche **230–231:** The Living **232–233:** Jsssjs Product Design + WickesWerks LLC + Yeadon Space Agency **234–235:** EJTECH **236–237:** yet|matilde s.n.c. **238:** Boris Miklautsch **239:** Julian Lienhard **240–241:** Giada Lagorio **242–243:** Trex Lab tech **244–245:** Shachar Fleischmann **246–247:** Spiber Inc. **248:** MPI of Colloids and Interfaces **250:** CSIRO **255:** NONdesigns, LLC **256:** Bioglow LLC **257:** John Horner **258:** Kevin Remy **259:** left Paul Bird; right Kevin Remy **260:** AnaElise Beckman, Alexandra Cohn, and Michael Zaiken **262:** NONdesigns, LLC **264:** top Pim Hendriksen; bottom Studio Roosegaarde **265:** Pim Hendriksen **266:** top row MIT Mediated Matter; bottom Steven Keating **267:** MIT MECH-2.009 **268–269.** Sheng-Ying Pao **270:** Nullohm AS **272–273:** NONdesigns, LLC **274:** Bioglow LLC **276–277:** John Horner **278–279:** Studio Roosegaarde **283:** MIT Tangible Media Group **284:** Burton Inc. **285:** The Self-Assembly Lab, MIT **286–287:** The Self-Assembly Lab, MIT **288:** Burton Inc. **290:** top Shadow Robot Company; bottom SynTouch **291:** SynTouch **292–293:** Branch Technology **294:** Felix Faire **296–297:** Squint/Opera **298–299:** François Lauginie **300–301:** Peter Tsai **302–303:** Blaine Brownell **304–305:** John Romanishin **306:** University of South Carolina **308–309:** KieranTimberlake

To Allison, Chris, Eloisa,
Mateo, and Ana Luz

Published by Princeton Architectural Press
A MCEVOY GROUP COMPANY

37 East 7th Street, New York, New York 10003
202 Warren Street, Hudson, New York 12534

www.papress.com

20 19 18 17 4 3 2 1 First edition

Editor: Jenny Florence Designer: Benjamin English

Special thanks to: Janet Behning, Nolan Boomer, Nicola Brower, Abby Bussel, Erin Cain, Tom Cho, Barbara Darko, Jan Cigliano Hartman, Lia Hunt, Mia Johnson, Valerie Kamen, Simone Kaplan-Senchak, Diane Levinson, Jennifer Lippert, Kristy Maier, Sara McKay, Eliana Miller, Jaime Nelson Noven, Esme Savage, Rob Shaeffer, Sara Stemen, Paul Wagner, and Joseph Weston of Princeton Architectural Press
– Kevin C. Lippert, publisher

LIBRARY OF CONGRESS
CATALOGING-IN-PUBLICATION DATA

Brownell, Blaine Erickson
Transmaterial next: a catalog of materials that redefine our future
ISBN 978-1-61689-560-0
1. Materials—Technological innovations.
2. Materials—Catalogs.
LCC TA403.6 .T153 2017 | DDC 620.1/1–dc23